Monographs on
Theoretical and Applied Genetics 11

Edited by
R. Frankel (Coordinating Editor), Bet-Dagan
M. Grossman, Urbana · H.F. Linskens, Nijmegen
P. Maliga, Oakland · R. Riley, London

Monographs on Theoretical and Applied Genetics

Volume 1 **Meiotic Configurations**
A Source of Information for Estimating
Genetic Parameters
By J. Sybenga

Volume 2 **Pollination Mechanisms, Reproduction and**
Plant Breeding
By R. Frankel and E. Galun

Volume 3 **Incompatibility in Angiosperms**
By D. de Nettancourt

Volume 4 **Gene Interactions in Development**
By L.I. Korochkin

Volume 5 **The Molecular Theory of Radiation Biology**
By K.H. Chadwick and H.P. Leenhouts

Volume 6 **Heterosis**
Reappraisal of Theory and Practice
Editor: R. Frankel

Volume 7 **Induced Mutations in Plant Breeding**
By W. Gottschalk and G. Wolff

Volume 8 **Protoplast Fusion**
Genetic Engineering in Higher Plants
By Y.Y. Gleba and K.M. Sytnik

Volume 9 **Petunia**
Editor: K.C. Sink

Volume 10 **Male Sterility in Higher Plants**
By M.L.H. Kaul

Volume 11 **Tree Breeding: Principles and Strategies**
By G. Namkoong, H.C. Kang, and J.S. Brouard

G. Namkoong   H.C. Kang
J.S. Brouard

Tree Breeding: Principles and Strategies

With 28 Illustrations

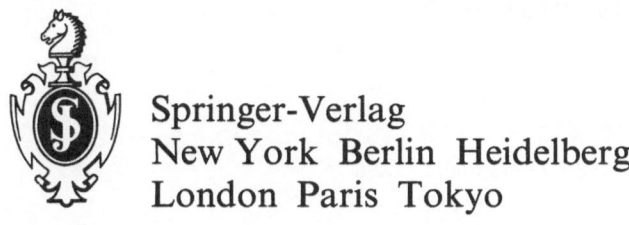

Springer-Verlag
New York Berlin Heidelberg
London Paris Tokyo

Gene Namkoong
Pioneer Research Geneticist
U.S. Department of Agriculture Forest Service
Southeastern Forest Experiment Station;
Professor, Department of Genetics
North Carolina State University
Raleigh, NC 27695-7614, USA

Hyun Chung Kang
USDA Forest Service
North Central Forest Experiment Station;
Department of Forestry
University of Wisconsin-Madison
Madison, WI 53706, USA

Jean Sébastien Brouard
International Resources Group Ltd.
1015 18th Street, NW
Suite 802
Washington, DC 20036, USA

Library of Congress Cataloging-in-Publication Data
Namkoong, Gene.
    Tree breeding : principles and strategies / Gene Namkoong, Hyun
  Chung Kang, Jean Sébastien Brouard.
      p. cm.—(Monographs on theoretical and applied genetics;
  11)
    Bibliography: p.
    Includes index.
    ISBN-13:978-1-4612-8392-8
    1. Trees—Breeding.   2. Forest genetics.   I. Kang, Hyun Chung.
  II. Brouard, Jean Sébastien.   III. Title.   IV. Series.
    SD339.5.N355 1988
  634.9′56—dc19   88-12352

Printed on acid-free paper

© 1988 by Springer-Verlag New York Inc.
Softcover reprint of the hardcover 1st edition 1988

Typeset by Asco Trade Typesetting Ltd., Hong Kong.

9 8 7 6 5 4 3 2 1

ISBN-13:978-1-4612-8392-8   e-ISBN-13:978-1-4612-3892-8
DOI:10.1007/978-1-4612-3892-8

Preface

It has become apparent, during discussions with students and colleagues in forest genetics, that a universal concern is the achievement of diverse goals of forestry from fiber production in industrial as well as farm forests to conserving forest ecosystems. Although we generally have several breeding methods available and several species to breed, we seek to satisfy multiple-use goals on diverse sites by management techniques that at best can only partially control edaphic environmental variation. The dominant approach, which was agriculturally motivated, has involved intensive effort with complicated breeding plans on single species for uniform adaptability and single-product plantations. However, this is obviously neither the only, nor necessarily the best, solution for the genetic management of tree species, and thus our intent in this volume is to develop ways to achieve multiple objectives in tree breeding. We include an array of breeding plans from simple iterated designs to sets of multiple populations capable of using gene actions for different traits in different environments for uncertain futures. The presentation is organized around the development of breeding from single- to multiple-option plans, from single to multiple traits, from single to multiple environments, and from single to multiple populations. However, it is not a complete "How To" book, and includes neither exercises nor instructions on data handling. It also does not include discussion of all modes of reproduction and inheritance encountered in plants.

A further motivation is to discuss what we the authors find particularly interesting about tree breeding, and this often focuses on problems in genetic theory and their implications in breeding practices. As such, this volume is not primarily intended for a broad plant breeding audience other than those in forestry, but we hope that it may be useful to other crop breeders. Although not primarily intended to be a textbook, we hope that it can serve as a study guide. It is also not intended to be a comprehensive tree breeder's handbook, but we certainly hope that it strongly influences tree breeding plans.

This volume is intended primarily for tree breeders, forest managers, and graduate students, and therefore we assume university training in forestry or applied biology, that is, a Bachelor's degree or several years experience in applied tree breeding. It is expected that the reader has some familiarity with basic genetics (e.g. Hartl, 1980), but we will cover the fundamentals of quantitative genetics. Falconer's (1981) text is a good introduction to quantitative genetics, while Namkoong's (1979) deals more specifically with forestry applications.

It is difficult to follow the theory or application of quantitative genetics without some knowledge of statistics, and we assume that readers can familiarize themselves with basic statistical theory and applied statistics by consulting any of the numerous textbooks in these fields, such as Mendenhall et al. (1981) for statistical theory and Steel and Torrie (1980) for applications. An understanding of univariate statistics is important, because we will extend the applications to multivariate cases. Multivariate statistical methods are described in Morrison (1976) and Johnson and Wichern (1982).

It is possible for breeders not interested in multiple breeding strategies to read only Chapters 1, 4, and 8, or for breeders not interested in breeding principles to read only Chapters 1, 4, 5, 6, 7, and 8.

Acknowledgments. The foundation thoughts to which we hope we have added some personal insights and the motivation for writing this book and exposing our concepts to further debate are due to our heritage of inspired teachers, disputative cohorts, and questioning students. With them we share a desire to use the science of forest genetics to better manage forested ecosystems, and to them we owe our deepest thanks for giving form to our intuition. We especially thank Sharon Friedman for reviewing and adding her insights to the book, Kjell Lännerholm for assistance in preparing the figures, Sheryl Brown Bolstad for preparing the index, and Lucy Nunnally for managing the production of the book starting with the earliest handwritten draft. The faults and errors of omission and commission, however, are our own.

GENE NAMKOONG
HYUN CHUNG KANG
JEAN SÉBASTIEN BROUARD

Contents

Preface   v

1   Tree Breeding Opportunities and Limitations . . .   1

1.1   Natural Selection   4
1.2   Gene Expression and Ecology in
      Selective Breeding   5
1.3   Relationship Between Natural and
      Artificial Selection   8
1.4   Conclusion   9

2   Defining Gene Effects   11

2.1   Models of Gene Action: One Locus   12
2.2   Models of Gene Action: Multiple Loci   14
2.3   Nonrandom Models   15
2.4   One-Locus Selection Models   16
2.5   Two-Locus Models   20
2.6 n-Locus Models   21
2.7   Selection, Inbreeding: Migration, Mutation,
      Population Size   25
2.8   Nonrandom Mating and Selection   31

3   Basic Concepts in Recurrent Selection   37

3.1   Changing Roles of Tree Breeding and
      Forest Genetics   37
3.2   Basic Model and Some Simple Models of
      Recurrent Selection   41
3.3   Gain and Limits to Selection   46
3.4   Developing Recurrent Selection Programs   50
3.5   Vegetative Propagation   53

4   General Recurrent Selection Systems   56

4.1   Mating Designs in Single-Population Breeding . .   57
4.2   Reciprocal Recurrent Selection for
      Hybrid Breeding   66
4.3   Multiple-Population Breeding   70

5   Selection Techniques   74

5.1   Heritability   74
5.2   Expected Gain from Direct Selection   77
5.3   Indirect Selection   79
5.4   Multitrait Selection Techniques   91
5.5   General Strategies for Multiple
       Breeding Objectives   94
5.6   Combining Gene Actions, Breeding Objectives,
       and Breeding Techniques   98

6   Environments   103

6.1   Genetic, Environmental, and
       Phenotypic Expression   103
6.2   Patterns of Response to
       Environmental Variation   105
6.3   Changes in Environments and Genotypes   111
6.4   Competition   115

**7   Coping with Present and Future
       Breeding Problems**   120

7.1   Coping with Inherent Breeding Problems   120
7.2   Strategies for Variable Ecological and
       Economic Environments   127
7.3   Multiple-Species Strategies   130

**8   Provenance Testing, Ecogeographic Surveys,
       and Conservation**   132

8.1   Species/Population Introduction and Testing . . .   132
8.2   Provenance Trials   134
8.3   Provenance Testing and Ecogeographic
       Surveys   148
8.4   Genetic Resource Conservation   149

References   161

Index   173

CHAPTER 1

Tree Breeding Opportunities and Limitations

Since the mid-1950s, tree breeders have demonstrated that simple selection-type breeding operations can substantially increase yield and modify many traits and adaptabilities. Yet, forest genetics has had only minor impact on the vast majority of commercial forest land and no impact on bringing marginally commercial land into production. Through experience we have learned that forest trees are often well endowed with genetic variation, but our inability to broadly affect most forestry practices indicates how little we have used all of the options of tree breeding. We have used intensive selection and breeding for highly intensive forestry management, but have not developed simple operations for low-input forestry. We generally breed for average, expected environments, for average functions of several traits, and for average gene effects, but do not develop options for multiple possible environments, for several sometimes conflicting trait values, and for complex gene actions. We have not developed the genetic options for the kinds of forestry practiced on most forest land.

Intensive tree breeding, as a tool of intensive forest technology, has extended agricultural plant breeding techniques to forest trees after making minor adjustments for plant size and generation time. It has been easy to adopt agricultural breeding systems as the paradigm for tree breeding while assuming that repeating simple selection cycles would generate optimal tree breeding population structures.

However, two problems require tree breeding to deviate from the simple agricultural pattern. First, the genetic management of agricultural crops themselves requires more elaborate structures of support populations than simple selection cycles from a single base population. For new pathogens, and changing environments or markets, the development of new breeding populations is complex, and requires generations of enhancement, testing, and breeding (Frey et al., 1984). Therefore, continued successful renewal of agricultural crops by such means is not at all assured, and for forest trees with less investment and longer generation times, the use of iterative breeding in a single closed population is even less likely to succeed.

Second, forestry is critically different from agriculture in the multiplicity of its products, in the large number of useful species, and in the heterogeneity and limited control of environments both now and in the future. Often several different products are derived from one species, and multiple products may even be derived from the same tree. Stands are also managed for multiple objectives, which may require breeding for different environmental conditions as well as for different products. At one extreme, the breeder may be exclusively concerned with intensive even-aged site management for a single species and a single product, but at the other may be expected to breed trees simultaneously for heterogeneous conditions and uses.

Many forest tree breeders also find themselves faced with a bewildering array of species and with breeding objectives that are based on unstated management objectives for future forests. Further, forests are almost always confined to those ecologically complex areas that adjoin field edges or agricultural borders and extend to the limits of vegetation zones. Most attention is paid to marginal agricultural land, but these areas may not be very large and may be always susceptible to agricultural conversion. Forest management and hence tree breeding must also anticipate temporal variations in biotic as well as physical environmental factors, both within a single crop generation and throughout the period of multiple breeding generations. Thus, with or without more intensive management of forest lands, the areas of concern will remain heterogeneous, and generally will be subject to a strictly limited capacity for environmental modification or control. The present intimate connection between natural and breeding populations with which forest geneticists are privileged to work may thus endure, and the research interests of evolutionists and breeders may continue to overlap strongly.

We are fortunate in that simple breeding works for simple traits in single environments because genetic variation is reasonably high in most species. However, when mixtures of gene actions exist in several traits with mixed values, and when trees must adapt to multiple environments that also include mixed values, an expansion of breeding concepts is required to formulate expanded breeding plans. The goal of this volume is to rationalize such expanded breeding plans when they become desirable or necessary. We contrast breeding for immediate gain with developing breeding programs that can respond quickly to changing economic and environmental requirements. We begin with the observation that selection generally does yield gain, but believe that forest tree breeding involves not only the breeding stock, but also the support populations and the provenance introductions needed to manage the genetics of many populations, Thus, breeding also includes conservation programs.

There are two necessary conditions for the success of breeding: (1) the existence of genetic variation in the trait of interest, and (2) the effectiveness of choosing parents for breeding. Early tree breeding techniques were generally based on collecting open-pollinated seeds from parents of good phenotype ("plus trees"). These seeds were then used directly for plantation establishment, or more commonly were established in progeny tests with a view to selecting those mother trees that produced the best offspring (in retrospective tests). Alternatively, because of the usually long generation time, the progeny themselves were rogued and used as a basis for generating further "improved" seed (in prospective testings).

Variants of these schemes arose with the possibility of grafting vegetative material from the plus trees into clonal seed orchards. The orchards could later be rogued on the basis of seedling performance in tests, while acting as a seed source of presumed superior genetic material in the interim. A summary of the types of early work in tree breeding can be found in Larsen (1951), Toda (1974), and Dorman (1976), while a more complete review of practical tree breeding activities can be found in Wright (1976) and Zobel and Talbert (1984).

In additon to intraspecific selection, interspecific hybrids have been fairly important in forestry (Conkle, 1970; Hyun, 1974; Fowler, 1978). Use of hybrids is facilitated when they can be propagated vegetatively, as for example with the Euramerican poplars in Canada (Zsuffa, 1979) or the hybrid eucalypts of the Congo (Delwaulle, 1985) and Brazil (Campinhos and Ikemori, 1977). Even when seed-based reproduction must be relied on, hybrids can prove useful in operational programs, particularly when attempting to cope with difficult environments; for example, *Pinus rigida* × *P. taeda* hybrids in Korea (Hyun, 1974). However, tree breeding programs have rarely been designed to take advantage of hybrid combining ability (Namkoong et al., 1980).

Many traits have responded, but selection may not always be very effective because some tree species, such as red pine, *Pinus resinosa*, are comparatively lacking in genetic variation (Fowler and Morris, 1977). Even when genetic variation does occur, the selected trait, such as early growth in longleaf pine, *P. palustris*, may be difficult to observe directly (Snyder, 1969) when observations are made on mature trees and when growth of older trees can only be estimated by observations on juvenile growth.

To select effectively, it may be tempting to consider the complexities of undirected evolution as inapplicable esoterica, but there are at least two reasons why we risk harm if we ignore evolution. First, there is a risk that we do in fact assume certain modes of genetic behavior, and if that assumption is based on simple views of the nature of inheritance which are false, then our actions are not likely to produce the results expected. We might, for example, assume that by the balancing effects of selection an optimally "fit" species composed of optimally "fit" populations has evolved, and hence that populations are evolving as the environment slowly shifts. From this assumption one might further extrapolate that local populations are always best and that breeders should not transgress local boundaries of ecological adaptation. Such an assumption, however, ignores the partial independence of traits immediately involved in natural versus artificial selection and the effects of mating pattern and evolutionary history on a current nonoptimal distribution of genotypes.

Selection in natural ecosystems is not unrelated to the objectives of artificial selection but neither are they totally congruent (Namkoong, 1985a), and even natural populations may respond only slowly to changing selective optima. In addition, patterns of selectively important environmental variables are generally independent of pollen and seed distributions and hence do not completely determine the patterns or even the existence of allelic variations (Gregorius and Namkoong, 1983; Namkoong and Gregorius, 1985). We discuss these issues further in Chapter 6.

The second risk is simply that we do not understand the effects of selection on the stability and persistence of populations. The observed patterns of variation may only be ephemeral stages in an evolutionary sequence that may ultimately result in different structures. On the other hand, what may appear to be trivial variations in population structure may be necessary for host populations to inhibit pathogen epidemics. It therefore seems the wiser course to accept this endowment of genetic variation and use it well, but to be not too profligate with the sources of that variation. We discuss objectives and plans for gene conservation and how it can be integrated with tree breeding in more detail in Chapter 8.

1.1 Natural Selection

It is obvious, from studies on electrophoretically detectable isoenzymes as well as from selection studies, that average levels of genetic variation in most populations of forest trees are very high (Hamrick, 1983). From these samples of gene loci, the percentage of all loci that are variable within species has averaged 68% for gymnosperms, and the frequency of alternate alleles is sufficiently high that trees would be expected to be heterozygous at 21% of their loci. Although the DNA content of some pine species is high, much of it may be in noncoding repeat sequence DNA (Hall et al., 1976), and hence the number of genetic loci may not be much greater than *Drosophila* and may be estimated as between 5,000 and 50,000 nuclear loci. An implication of these figures is that differences among individual trees, among stands, and among species are based on a huge number of gene differences; variation is not necessarily based only on a few key genes that differentiate individuals, a few more that differentiate populations, and yet a few more that differentiate species.

However, it is not at all clear if such extensive variation exists for all variable loci or if certain classes of genes contain more variation than others. It does seem clear that certain base-pair sequences are more variable, that certain regions of genes are more likely than others to contain variations, and that these variations are generally not associated with known functional differences (Kimura, 1983). Therefore, some of the observed variation may have no selective significance at all, whereas some may have subtle effects that are not readily associated with easily observed selective effects. Whether selectively neutral or currently active, however, the genetic variation that does affect traits can be expected to involve multiple gene effects that are expressed through physiological and developmental processes which are characteristically controlled by various feedback mechanisms and may confer net advantage or disadvantage in complex ecological systems. That is, it will rarely be true that much of the observed variation in genetic material will be associated with single genes of large, simple effect on "fitness."

Observations of genetic variation and response to selection, which are often great in forest trees, indicate that high levels of variation commonly continue to exist regardless of how such variation evolved. For genes having a neutral effect on viability, variation associated with reproductive behavior can cause patterns of variation that are indistinguishable from patterns caused by viability selection. In fact, reproductive biology has a substantial impact on patterns of variation, and its effects are often confounded with inherited growth and other traits that affect viability. Thus, the extensive genetic variation within coastal populations of Douglas-fir and loblolly pine (Yeh, 1981; Roberds and Conkle, 1984) may be associated with an extensive pattern of wide intercrossing. Sharper population distinctions among populations of yellow poplar, *Liriodendron tulipifera*, may be associated with much more limited outbreeding (Brotschol et al., 1986). There is a general association of life history traits with patterns of intra- and interpopulation discrimination in allele frequencies (Hamrick, 1983), and a general tendency for the more boreal species to exhibit wider intercrossing and less population differentiation than do tropical species (Stern and Roche, 1974; O'Malley and Bawa, 1987). However, these patterns are not uniform, are subject to temporal fluctuations, and may only be true in a very broad sense (Namkoong, 1984a); they also depend on the species, the populations sampled, and the traits under study (Namkoong, 1985a). Thus, patterns of trait variation resulting from neutral or reproductively associated effects are not easily separated from more complicated types of selection on growth or viability traits, but may nevertheless be the cause of observed distributions of allelic variation.

On the other hand, natural selection on traits can also generate a wide variety of allelic distribution patterns, and because many traits readily respond to artificial selection, we can infer the possibility that natural selection could also have produced observed trait variations. A reasonable hypothesis on the existence of such inherited clinal trends (such as the growth response of *Picea abies* in Sweden) is that they reflect clinal trends in environmental stresses (Langlet, 1963). Kellison (1966; 1970) found significant correlations between some leaf and fruit morphological traits and environmental variables in natural populations of yellow poplar. Local population differences have also been shown to reflect local patches of environmental stress, such as tolerance to heavy metal toxicity in grasses (Jain and Bradshaw, 1966; Antonovics, 1968). If taken to the extreme of assuming additivity of gene effects within and between loci in addition to consistent selection pressure and wide migration, then genetic variation in growth response would be expected to disappear in such populations. Since variations in viability traits persist, at least one of these assumptions is invalid. Hence, the existence of generally high levels of genetic variation has generated studies on how gene effects, or variable environments, can maintain genetic variation and how population size affects patterns of variation. While these issues are too complex to adequately discuss here, interaction of gene effects, spatial and temporal variations in environments, and finite population sizes are sufficient to make the discernment of pure causes exceedingly difficult and perhaps meaningless.

1.2 Gene Expression and Ecology in Selective Breeding

The effects of natural selection on specific traits or genes are difficult to observe and reproductive patterns can mask selective effects, but the effects of consistent artificial selection reinforced by restrictive mating in recurrent cycles are much more clearly seen. So long as trait differences can be observed and are heritable, then simply limiting reproduction to appropriate parents can achieve gain. In the long term, repeated breeding can accumulate short-term gains. Although there would be differences in long-term responses among trait types such as growth, pest and disease resistances, fiber composition, and proportion traits, long-term selection experiments have shown that accumulated gains are possible. Continued progress is observed in selection experiments in fruit flies (Hill, 1982) and maize (Dudley, 1977) and serve as models for the probable behavior of forest trees in response to repeated cycles of selection. In each generation, gains are made by selecting parents, and as long as heritable variation continues to exist, gain is accumulated (Allard, 1960). It has taken a few generations, however, for the mean of advanced generations to approach the extremes of the original parental populations and for later generations to produce phenotypes never before seen in the wild populations (Allard, 1960).

Thus these examples from nontree species strongly indicate that simple breeding operations can achieve short- and long-term gains even with poorly heritable traits.

These positive results, however, mask two critical factors that influence success in tree breeding. First, the observed gain from selection roughly fits the simple gene action models such as single-locus models widely used in quantitative genetics. Therefore, tree breeders are not encouraged to clearly distinguish between the true nature of gene expression and their models of gene action. Second, the progress from selection demonstrated in these examples was obtained under well-controlled, repeatable environmental conditions. This is certainly not true in tree breeding.

In the following section, we further address these two issues. In Chapter 2 we discuss in more detail those models of gene action that are the basis for selection and breeding theory, but here we merely assert that our observations of selection response are based on gene effects that may vary. In Chapters 7 and 8 we discuss breeding and conservation under changing uncertain environments.

1.2.1 Gene Expression

Obviously, different gene effects are responsible for the response of different traits, and some of these are easier to handle in a breeding program than others. It is also obvious that if we knew that certain identifiable genes affected certain traits and also knew how they worked, we could be more efficient in genetic management. This may well be possible for some traits with particularly simple developmental pathways and for genes that can be readily identified as causal agents. Thus molecular genetics can be expected to have its most immediate and direct impact on breeding when simple organisms or major genes with well known and simple developmental effects are involved.

However, many genes for most traits of economic or ecological significance operate either in concert with other loci or in fairly complex developmental pathways within the cell, the organ, or the entire organism, where regulatory mechanisms affect complex trait expressions. Therefore, single-gene manipulation will not generally be an efficient means of breeding. We may then ask about the relationship between genes and the traits that are affected by breeding. If we define a "quantitative" trait as one with many genes affecting it, each of small effect, then it is actually a model for single gene–single trait effects simply added over many loci. While this will be used in later chapters as a convenient concept for the short-term effects of selection on genes, it is clearly a naive concept of gene action.

Consider, for example, certain traits that may be heritable, such as the quality of wood for making the body of guitars. The value of wood for this purpose depends on many elements including fiber length, density, growth rate, and length of clear wood, all of which are heritable. There is little question that, if so desired, a silvicultural and breeding system could improve the quality and quantity of that kind of wood, and although its heritability might be low, continued selection response can reasonably be expected. Yet there can also be little question that no such gene as a "guitar-quality" gene has ever or will ever exist.

Indeed there is no such gene as a gene for density or growth rate, although there are genes that affect all of those traits. The danger of reification is that while such concepts are useful as approximations, we may forget that genes affect processes and response functions and are not independently observed outside of the environment within which those genes and organisms operate. Thus, gene effects are real but are not to be confused with their models, and the effects of genes integrated within individual organisms are complex and require clarification. In general, then, selection does not act directly on genes; rather, selection effects are mediated by genes within certain environmental contexts. Quantitative trait loci may not exist as identifiable entities, even though loci that affect quantitative traits do.

We thus consider that many genes affect traits and that selection affects many loci, but must then ask which traits are to be selected. In forest tree breeding in particular, with its imprecisely controlled environmental variations over space and time, a major problem arises when choosing traits and environments for which breeding can be most valuable. Opportunities clearly exist to affect many traits, and indeed radical new products and behavioral responses can be bred that could substantially affect forestry operations. To be effective, however, the choice must be made, and a finite set of breeding methods, traits, and production sites chosen for breeding operations. In Chapters 3 and 4 we discuss breeding methods, and in Chapter 5 the analysis of different trait inheritances based on simple genetic models. Then, in Chapter 6, we consider environmental effects and the simultaneous effects of choosing breeding methods for sets of traits in sets of environments. In these chapters, our approach is to consider the simplest condition first and then to expand to more complex plans. Thus, we first consider fixed-gene effects for single traits for adaptation to a uniform environment before considering when more complicated plans become optimal.

1.2.2 Selection in an Ecological Context

In addition to breeding for genetic effects on processes within organisms, we also consider breeding to affect interorganismal processes. Because forest trees grow in stands and because the economic and ecological products we value are group properties, any group effects that are not merely accumulations of independent individual effects may require breeding for group objectives. Thus, competitive and mutualistic effects of trees on other individuals of the same species, other tree species, other vegetation, and populations of other organisms internal and external to the tree, can all be objects of selection. For example, the ability to grow in stands may not be the same as the ability to grow in the open, or there may be differences in growth response to stand density. In this case, selective breeding may require testing for response to a range of densities under specific silvicultural regimes (Nance and Wells, 1981). In addition, density alone may not account for all the intergenotypic competitive effects, and hence an optimum breeding program may not be easy to develop without substantial further study of competitive growth phenomena. Such study would require interdisciplinary research by geneticists, physiologists, and silviculturists. At the present stage of tree breeding, we usually select on individual tree performances that have uncertain effects on the populations undergoing manipulation and on the ecosystems of which these populations will be a component. In such situations, the strategy we follow is to choose selection goals as wisely as possible given present information, which might require selecting for broad adaptability, but also to continue research and development programs to guide future correction. Selecting for broad adaptability can be at the individual, population, or species level. This is a strategy question we address in Chapters 4–8.

The difficulty in choosing selection goals is also apparent in breeding for resistance to insects and fungi. The agricultural models of selecting for sequences of specific, "vertical" resistances with continued, rapid adaptation by the pest population would be difficult to follow in forestry, and may not even be optimal for agricultural crops themselves. To control pests in only partially managed forest ecosystems clearly requires substantial research in genetics, pathology, entomology, and ecology but currently breeders must operate with a substantial degree of ignorance. Hence the optimal solutions to breeding forest trees for partially managed ecosystems are not simple imitations of historical agricultural systems, nor are there any obvious specific alternatives. For some intensively managed forests, breeding may simplify to more quasi-agricultural systems, but for many a broader vision is required that incorporates breeding within an ecological context.

This may also require us to develop a structure to the genetic diversity that might imitate nature more than agriculture. The tree breeder must consider a broad range of forest management systems, and for each species, her effect on forest productivity may depend on how well selective breeding is adapted to the types and intensity of management practiced. This may require intensive seed orchard operations for uniform industrial plantations, or extensive seed orchards for direct seeding mixed varieties, or merely ensuring that genetic diversity is enhanced in "natural" regeneration. The breeder can also exert wide influence over the genetic variation of future forests by reducing breeding costs and making improved populations available for the less capital-intensive forms of forest management.

1.3 Relationship Between Natural and Artificial Selection

In Sections 1.1 and 1.2, we discussed natural selection and the basis of artificial selection of forest trees as separate functions. We may wish to consider whether very sharp distinctions necessarily exist between natural and artificial selection, and whether we may draw any lessons about breeding populations from natural populations. It is clear that, by selection and forced limited mating, a deliberate directional change in genes and traits is feasible. But by breeding on one set of organisms, we do affect other elements of the ecosystem both now and in the future. This fact alone need not prevent breeders from pursuing short-term objectives, but it should dictate that a more global view be taken of our effects on forest ecosystems including, but not restricted to, the future of the populations we are directly affecting by our breeding.

Therefore, a persistent problem is the necessity of operating with both immediate and long-range objectives and with both local and global goals. Short-term objectives may include increasing growth and survival of timber trees or pulp yield for paper and cellulose products on specifically managed forest lands. Long-term objectives may include progressively improving the yield of several products for a range of possible forest sites, as well as maintaining and developing resistance to evolving insects and pathogens. These objectives may require that simple short-term breeding plans be supplemented or modified in other structures of breeding populations to ensure progression to an optimum structure. In Chapter 7, we concentrate on the problem of achieving these objectives and on the relationship among breeding plans for multiple objectives.

One problem of breeding to ensure progressive adaptability to changing environments lies in structuring populations for variations that can only be partially controlled by management interventions and predicted with high uncertainty. This may require breeding for wide adaptability within single breeding populations, or among deliberately constructed sets of differently adapted populations, or both. Thus, breeding may create more diverse population structures than presently exist, and may ensure greater population stability in terms of evolutionary potential and economic yield. In fact, population stability may be one goal of breeding or controlled reproduction, especially for those who consider breeding to be a form of guided evolution. On the other hand, a goal of stability may be incorporated at the species or other level of organization, if it is a goal at all. Other goals may be better described as maximization of some measure of expected yield with discounting of future benefits, while other goals may merely be maintenance of populations capable of responding to local, immediate objectives of ecosystem resilience but without an objective of cumulative or progressive improvements.

These alternate goals for future populations are similar to debates and concepts about the existence of fitness maximization in natural evolution, but it is not necessary that one's future goals coincide with beliefs about past evolution. Natural evolution has been viewed by some as advancing to some stable, optimum equilibria, and by others as responding to local selection with only slight delays caused by mating limitations; others consider evolution to be usually static, with random changes in gene frequencies of little consequence in any directional shifts.

One may choose to believe in the natural progress of evolution and that breeding can redirect this progression to human economic ends. Alternatively, one can believe in stasis in both evolution and breeding, and breed only for short-term trait changes but maintain large unimproved, more or less natural populations. However, it is also possible to recognize evolutionary progression but believe that breeding is best directed to short-term objectives, or, conversely, to believe that evolution is static but that breeding can nevertheless be directed into long-term progressive programs. It is desirable to know the current status of population variation and the evolutionary history of the population, whether the goal is to maintain or to change it. We discuss studies of natural population for these purposes as well as for immediate breeding uses in Chapter 8.

1.4 Conclusion

The thesis of this volume is that forest tree breeding can change traits. The natural endowment of genetic variation in forest trees is high, as might be expected for species not previously subjected to directed breeding efforts. This variation, while distributed in patterns that are not well known or consistent, is easily exploited.

However, when such factors as population size, restraint on trait selection, and environmental specificity and control are not considered, severe problems can soon develop in simple selective breeding operations. The very extensiveness of forests, the number of forest species and products, the variability of forest environments, and the uncertainty of future ecological and economic demands all require more careful planning for future breeding populations than is required for agricultural crops. Tree breeders cannot often afford to concentrate efforts on developing a few elite lines or populations in a few, globally adapted species.

Because these factors also affect the utility of other products of our breeding, we must consider more global goals as well as the short-term objectives of breeding. Although these long-term and global goals may require that we impose a structure on breeding populations, we must start by considering the simplest genetic models and breeding plans. If the simple models provide a sufficiently good approximation to reality, then simple breeding plans would be sufficient for achieving those goals. For those genes, traits, environments, and species that depart from the simple models, we then build alternative breeding plans and suggest research and development programs that allow the breeder to bridge between the simple and the complex. The simple breeding methods, if conducted efficiently and precisely, can generate economic improvement with many more species than are currently included in breeding programs, and still provide safeguards for future genetic variations. Armed with an array of breeding program options, the breeder can broaden the influence she has over the potential yield and diversity in the genetic structure of forests for all of its products and uses.

CHAPTER 2
Defining Gene Effects

Current developments in molecular genetics are rapidly overtaking our concepts of gene structure and action and of their control of the development of organisms. The structure of genes and even their location are now found to be far less simple and stable than previously thought, and the events genes affect in the maintenance of cells and cell products are also subject to control directly by other gene actions and indirectly through the genes' own feedback mechanisms.

It is also known that DNA in cellular organelles such as mitochondria and chloroplasts can act in similarly complex ways and, with nuclear genes, can interactively affect trait expressions. Thus, it is obvious that the effect of a gene is often modified by other genes even within the cell; this in itself can produce nonsteady states of gene products even if each gene in a single pathway has simple effects. At the tissue level of observation, many gene effects are embedded within developmental systems, which themselves involve multiple cells and processes that interact in far from simple dynamics. Hence, simple direct gene effects are rarely possible to track to their expression at the level of the organ or whole organism.

However, although there are few simple, direct effects of genes on complex traits and developmental pathways, single gene effects can be identified; genes can be modified, transferred, and expressed in exotic genomes. This is possible when genetic defects exist in otherwise satisfactorily operating genomes that can be compared in either their gene product or gene structure to those of "normal" genes. Familiarity with normal enzymes, which have known effects compared with nonnormal, permits a pathway to be constructed and the originating defect in the nonnormal gene identified. Thus, even if the developmental pathway is complex, single genes of major effect can be identified and changed. A second way to use direct gene effects is to survey processes that occur in all organisms or to conceive, de novo, of processes which may possibly occur. If sufficiently simple pathways exist that can be "added on" with minor effect on other genomic functions or can modify simple functions, then gene substitutions or insertions will have predictable effects.

A major biological question at this time, therefore, is the extent to which more or less independent clusters of development exist, particularly in simply organized systems. For many traits, the development of expression is not simple. During the life of a tree, as gene expressions are accumulated or are turned on and off, gene effects may change and other epigenetic phenomena may assume greater importance in affecting fitness. Genes that moderate the interactions of reproductive organs with vegetative tissues and thus produce a reproductive schedule are determiners of fitness although far removed from direct gene products. Even more remote are genes that affect interplant cooperative or competitive interactions.

Obviously, sexual reproduction requires considering gene effects on fitness as functions of other individuals and sexes, and that there are various forms of detrimental and beneficial effects of individual or stand interactions indicates that the ultimate effect of genes is a function of the genes in other individuals. Genes that endow resistance, for example, depend on virulent organisms so that they may be recognized as possessing resistance and may endow susceptibility for other types of virulence.

Thus many genes are likely to affect genotypic expression for traits of any developmental complexity, and any individual gene involved in developmental processes is likely to have effects on several component trait expressions. Thus it seems at best naive, and possibly even dangerous, to adopt the view of Dawkins (1982), embodied in the concept of the selfish gene, that gene level effects drive evolution and that individuals are merely the harboring encasement for genes. Gene effects in large part are simply not meaningfully understood if limited to their direct products, and if they are functions of many other genes and effects at the individual level and even interindividual level, it is meaningless to focus on autonomous genes as the objectives or goals of evolution or breeding. It is illusory to focus attention on merely finding the right genes, by either breeding or molecular manipulation, and inserting or collecting them into individuals in order to design better crops. Nevertheless, single genes can affect resistances such as those against Tobacco Mosaic Virus, and can affect simple but economically critical traits. Adding such effects to trees bred for all of the more complexly inherited traits can be a useful supplement to, but cannot substitute for, breeding.

2.1 Models of Gene Action: One Locus

In spite of the general remoteness of direct gene action from phenotypic expression, and the dangers of reifying genes for traits, it cannot be denied that many traits and reactions are inherited and that selective breeding is often effective. At least in the local sense for a given set of genes, in a given set of genomes, and in a given set of environments, genes can be reasonably modeled as having at least a small constant effect when averaged over a limited set of genetic and environmental variables. Thus, a shift in the rate constants of an enzyme effect caused by an allelic change may, within this context, have an average effect on a trait expression in trees. In this sense, there are genes that affect such traits as height growth of trees, even if there is no such thing as a gene for height growth.

Further, if genotypes differ in their average trait expression within a set of environments, we can then define an average effect of a gene substitution in a population as a weighted average of phenotypes, as follows:

Let $Y_{A_1 A_1}$ = Average phenotypic value of all individuals of genotype $A_1 A_1$

$Y_{A_1 A_2}$ = Average phenotype of $A_1 A_2$ individual

$Y_{A_2 A_2}$ = Average phenotype of $A_2 A_2$ individual   (2.1)

for two alleles at locus A. Then the substitution of A_1 for A_2 has two effects depending on whether $A_1 A_1$ is compared with $A_1 A_2$ or if $A_1 A_2$ is compared with $A_2 A_2$. In this sense, the frequencies of the genotypes help determine the strength of the effect in the population. Assuming an allele frequency of p, the genotypic frequencies following random allele associations of the $A_1 A_1$, $A_1 A_2$, and $A_2 A_2$ trees are p^2, $2p(1 - p)$, and $(1 - p)^2$, respectively. Then, an average phenotype, $\bar{Y}$, can be defined as the weighted average:

$$\bar{Y} = p^2 Y_{A_1 A_1} + 2p(1 - p) Y_{A_1 A_2} + (1 - p)^2 Y_{A_2 A_2} \tag{2.2}$$

Similarly, an average phenotypic effect of each allele can be defined by weighting the effect of each phenotype by the frequency at which it occurs with the allele in question. This same $\bar{Y}$ then is described as a function of allele effects and allele frequencies. We can then describe population means and variances in terms of alleles rather than phenotypes and can then analyze the phenotypic evolution of populations in terms of genetic effects. For allele A_1, its effect is $Y_{A_1 A_1}$ whenever it is combined with A_1, which occurs with frequency of p, and its effect is $Y_{A_1 A_2}$ when it is combined with A_2, with frequency $1 - p$. Thus, the average phenotypic effect of allele A_1 is Y_{A_1}, and can be written as:

$$Y_{A_1} = p Y_{A_1 A_1} + (1 - p) Y_{A_1 A_2}$$

Similarly,

$$Y_{A_2} = p Y_{A_1 A_2} + (1 - p) Y_{A_2 A_2} \tag{2.3}$$

and we can then define the average genotype $\bar{Y}$ as a weighted average of average allele effects as:

$$\bar{Y} = p Y_{A_1} + (1 - p) Y_{A_2}$$
$$= p^2 Y_{A_1 A_1} + 2p(1 - p) Y_{A_1 A_2} + (1 - p)^2 Y_{A_2 A_2} \tag{2.4}$$

Although it may be possible to define gene effects at the molecular level by some external standard of activity, allelic effects can only be defined relative to other allelic effects at the phenotypic level. Thus, by scaling these $Y_{A_i A_j}$ effects around zero, and using just two parameters, we can without loss of generality allow $Y_{A_1 A_1} - Y_{A_2 A_2} = 2u$, by assigning $Y_{A_1 A_1} = u$, and $Y_{A_2 A_2} = -u$, and allow the heterozygote to be $Y_{A_1 A_2} = au$, for any real-valued a. If $a = 0$, then no genetic dominance exists, and $a > 1$ or $a < -1$ implies overdominance of one of the alleles. In this notation, $Y_{A_1} = u[p + a(1 - p)]$, and $Y_{A_2} = -u[(1 - p) - ap]$, the so-called average effects α_1 and α_2, respectively.

We can also derive from these definitions that the variance in the average effects of alleles, that is the variance of Y_{A_i}, is

$$V(Y_A) = p(1 - p)(Y_{A_1} - Y_{A_2})^2$$
$$= p(1 - p)u^2[1 + a(1 - 2p)]^2 \tag{2.5}$$

The total variance among genotypes is

$$V(\text{Genotypes}) = p(1 - p)u^2[1 + a(1 - 2p)]^2 + 4p^2(1 - p)^2 a^2 u^2$$
$$= \text{additive variance} + \text{dominance variance} \tag{2.6}$$

Thus, the total genotypic variance is the sum of the variance of the average effects of alleles, plus a portion that Comstock and Robinson (1948) termed the dominance genetic variance, while the variance of average effects is called the additive genetic variance. Other parameterizations such as Falconer's (1981) are equivalent.

It is obvious from these definitions and the statistics derived from them that these gene effects are only defined within a populational context. Genetic variances are defined for specific gene frequencies which, by assumed random association, generate Hardy–Weinberg genotypic frequencies. Thus, all definitions are consistent for fixed frequencies. As is obvious from Section 2.1, we also are restricted to some understood genomic background over which the marginal effect of genes can be measured. Thus, while complex gene interactions may exist, as long as genotypic differences can be measured an effect can be defined, and the genotypic differences are presumed to be approximately consistent over some presumed variations in the genomic background.

We proceed in this chapter with these assumptions, but must remember that this approximation is a convenient assumption. In later chapters we also discuss the assumed external environmental variations over which these marginal allelic effects are measured. Suffice it to say here that we assume consistent gene effects for some set of environments, but recognize that dangerous reifications can result from assuming that gene effects can always be separated from environmental effects in simple linear models (Lewontin, 1974).

2.2 Models of Gene Action: Multiple Loci

Defining gene effects at a locus by homozygote differences and heterozygote deviations from the average of their homozygotes, we assume initially that we can at least theoretically measure such effects over some array of other gene effects; further, although those other loci may vary from tree to tree, we assume that average genotypic differences would exist. We now extend our model to consider a second locus that might also be defined as having a marginal effect on the same trait, and assume that it too can have effects and variances attributed to it in exactly comparable ways.

The question of concern is, how do these loci jointly affect the trait and how can we describe and measure such joint actions? If we first assume an independence of action, we might define that as having the same effect on the trait regardless of the actions and allelic states at other loci. This is most commonly taken to mean a linear independence, as in Cockerham's (1954) models, such that the effect at a locus is added to that at other loci and is the same regardless of other locus effects. Thus, a locus may have a difference between two homozygotes of say 20 units, and regardless of whether other loci contribute 200 or 2,000 units to mean performance, the difference of 20 units remains constant.

This is the linear sense of independence we use in order to keep this book aligned with other texts on quantitative genetics, but it is neither the only nor possibly the most ultimately useful concept to follow. For example, one can conceive of a gene operating in a tandem sequence and having a multiplicative effect on other gene products independent of the particular allelic state at the other loci. A multiplicative independence or some other functional independence may be defined in such cases and variances resulting from such effects also described. However, we leave such models for other developments.

If we consider the "main effects" of marginal gene effects as the main objects of study, then it is natural to expand the linear model for genotypic value G, to include various interlocus interaction effects. Thus, we can build three types of interaction between the main additive and dominance effects for locus A (with alleles i and j), and locus B (with k and l alleles):

$$
\begin{aligned}
G_{A_i A_j B_k B_l E} = {} & \mu + \alpha_{A_i} + \alpha_{A_j} + \delta_{A_i A_j} + \beta_{B_k} + \beta_{B_l} + \delta_{B_k B_l} \\
& + \sum_{i,k} \alpha_{A_i} \beta_{B_k} + \sum_{i,k,l} \alpha_{A_i} \delta_{B_k B_l} + \sum_{i,j,k} \beta_{B_k} \delta_{A_i A_j} \\
& + \delta_{A_i A_j} \delta_{B_k B_l} + \varepsilon_{ijklm}
\end{aligned}
\tag{2.7}
$$

where the $\alpha\beta$ effects can be defined as additive-by-additive epistasis, the $\alpha\delta$ and $\beta\delta$ as the additive-by-dominance epistasis, and $\delta\delta$ as dominance-by-dominance. These effects can be defined (Cockerham, 1954) independently, and hence the summation of their variances exhausts all the genetically associated variation. As with independent single-locus effects in which the additional dominance variances could be summed, not only can the single-locus effects be summed, but all two-locus effects also can be summed. Furthermore, the model can be extended to include more loci with single-locus, two-locus, and up to n-locus effects, and can exhaust all the genetically associated variation. These variances are given by Cockerham (1954) and by Namkoong (1979) in terms of phenotypic effects.

2.3 Nonrandom Models

In the derivation of these definitions of effects and variance, it was noted that these effects are measured in populations with particular gene frequencies. The definitions of effects, which strongly relate to frequency-independent concepts of gene action, involve the frequency of alleles and genotypes. Both the average effects and the variances are functions of genotypic effects and frequencies, and hence both can change if either gene or genotypic frequencies change. In particular, two types of effects can alter the measures discussed: inbreeding and linkage disequilibrium. The first effect engenders deviations from Hardy–Weinberg frequencies of genotypes at a locus, and thus the phenotypic effects of the various genotypes are not weighted by their frequency of occurrence in the stated equations. The second effect engenders deviations of two-locus combinations of genotypes from the frequencies that are expected under the assumptions of independence, and hence the phenotypic effects are not weighted according to the frequency of their occurrence.

We note first that without inbreeding or general nonrandom association of alleles, the variance is simply the sum of variances where σ_α^2 and σ_β^2 are the additive variances, and $\sigma_{\delta A}^2$ and $\sigma_{\delta B}^2$ are the dominance variances at locus A and B:

$$
Var(\text{Genetic}) = 2\sigma_\alpha^2 + \sigma_{\delta A}^2 + 2\sigma_\beta^2 + \sigma_{\delta B}^2
\tag{2.8}
$$

If, however, there is a nonrandom association of alleles, then the genetic variance would have to include some of the implied covariances, which are induced by frequency effects and not by the direct effect of alleles:

$$Var(G) = 2\sigma_\alpha^2 + \sigma_{\delta A}^2 + 2\,Cov(\alpha_i, \alpha_j) + 2\,Cov(\alpha_i, \delta_{ij}) + 2\,Cov(\alpha_j, \delta_{ij}) + 2\sigma_\beta^2$$
$$+ \sigma_{\delta B}^2 + 2\,Cov(\beta_k, \beta_l) + 2\,Cov(\beta_k, \delta_{kl}) + 2\,Cov(\beta_l, \delta_{kl}) \qquad (2.9)$$

If we model the nonrandom association that may result from any of several forms of nonrandom mating including inbreeding, as Wright (1922) did, then the frequencies of genotypes can be modeled as departures from expected or random-model frequencies.

For multiple loci, the addition of effects and variances over independent sets of loci is still interpretable, but because the effects at each locus are dependent on genotypic frequencies, any inbreeding causes interpretive problems. Various kinds of departures from random mating exist in forest trees (Namkoong, 1965), and the relative sizes of estimated genetic variances under inbreeding are often unpredictable.

If multiple loci are considered, nonrandom association of allele frequencies between loci can occur by selection or sampling effects even without inbreeding, and even without physical linkage of loci, associations can linger in populations through subsequent random mating generations (Cockerham and Weir, 1983). If the frequency of AB gametes, f_{AB}, is not equal to the product of the allele frequencies, $f_A \cdot f_B$, then the disequilibrium, $(f_{AB} - f_A \cdot f_B)$, is not zero. For example, if hybrids are generated by crossing among formerly isolated populations or species, several alleles enter the new population together from each parent population. The pairwise associations are dissipated by subsequent random mating within the hybrid population, but a disequilibrium of $(1 - r)^n$ remains, where r is the recombination rate and n is the number of generations.

Interestingly, it can be readily derived that each locus immediately returns to Hardy–Weinberg equilibrium frequencies if gametes randomly associate even as the interlocus, nonrandom associations dissipate more slowly. This may be a source of the existence of disequilibria in southern pine beetle populations in which past interpopulation hybridizations may have lingering effects in addition to nonrandom mating and selection (Roberds et al., 1987). However, in these as well as single-locus cases, selection effects must also be considered as capable of inducing changes and departures from equilibrium conditions.

2.4 One-Locus Selection Models

When selecting at the diploid phenotypic level under either natural conditions or human direction, we assume that marginal effects of genes are present and constant within some genomic and environmental set. By this marginal effect, we do not necessarily mean that the effect is a scalar constant; as we shall see, it may be a function of other variables. Considering simple effects for each genotype, we can limit our model to a phenotypic effect that affects the probability of the genotype achieving adulthood and reproducing. Assuming the independence of individual selections, and sufficiently large populations that the frequencies of survival and reproduction are as expected by the model, a discrete generation change in gene frequency can be readily derived. For this situation, two selection parameters are needed to describe the selection effects on three genotypes:

Genotype	Initial frequencies	Relative selective values	Frequencies after selection
$A_1 A_1$	p^2	$1 - s$	$p^2(1 - s)$
$A_1 A_2$	$2p(1 - p)$	1	$2p(1 - p)$
$A_2 A_2$	$(1 - p)^2$	$1 - t$	$(1 - p)^2(1 - t)$

In this case $1 \geq s$, t, and if both are less than zero, underdominance is said to exist. If both are greater than zero, overdominance exists and if s and t are of different sign, directional selection exists. Assuming random association of gametes following selection, the changes in frequency of the A_1 allele as a proportion of the changes in all alleles can be derived. The new frequency of A_1 is p' and is

$$p' = \frac{p^2(1 - s) + p(1 - p)}{p^2(1 - s) + 2p(1 - p) + (1 - p)^2(1 - t)} = \frac{p(1 - sp)}{1 - sp^2 - t(1 - p)^2} \qquad (2.10)$$

and the change in the A_1 allele frequency Δp is

$$\Delta p = p' - p = \frac{-p(1 - p)[sp - (1 - p)t]}{1 - sp^2 - t(1 - p)^2} \qquad (2.11)$$

If $sp > t(1 - p)$, then the A_1 allele frequency declines because the denominator is always positive and $0 \leq p \leq 1$, but if $sp > t(1 - p)$, it increases. However, if $sp = (1 - p)t$, $\Delta p = 0$ at $p = t/(s + t)$. This can only occur if s and t are both positive or both negative. If both are negative, the equilibrium is unstable, and any positive deviation from $p = t/(s + t)$ results in $\Delta p > 0$, and vice versa, thus driving the allele frequency to either 0 or 1. In single-locus models, Ginzburg (1983) showed that with multiple alleles only an average heterozygote superiority over all allele combinations is required, and that this is only a necessary and not a sufficient condition for a polymorphism to be stable.

In continuous models of selection, the generations are not considered to be discrete, and hence changes in gene frequencies and selection effects are not imposed on the whole population at the same time. The additional approximation imposed in continuous rather than discrete generation models is that the changes in gene frequencies are sufficiently slow that ordinary differential equations can model the dynamics. Such an approximation carries some internal inconsistencies and constraints on types of selection allowed (Gregorius, 1984) but nevertheless is useful for simplified dynamics. For a model similar to that just described, we derive changes in allele frequencies from the selective values of the individual genotypes. We define a selective value or "fitness" of genotypes as a net per individual replacement rate that may be scalar constants or functions such as: $\tau_{A_1 A_1}$, $\tau_{A_1 A_2}$, $\tau_{A_2 A_2}$.

As before, we also assume random mating and Hardy–Weinberg proportions and define allele values or fitnesses as

$$\tau_{\bar{A}_1} = p\tau_{A_1 A_1} + (1 - p)\tau_{A_1 A_2}$$

$$\tau_{\bar{A}_2} = p\tau_{A_1 A_2} + (1 - p)\tau_{A_2 A_2}$$

and

$$\bar{\tau} = p\tau_{\bar{A}_1} + (1 - p)\tau_{\bar{A}_2} \tag{2.12}$$

for a two-allele model.

From these definitions, and considering that

$$p = \frac{N_{A_1 A_1} + 1/2(N_{A_1 A_2})}{N_{A_1 A_1} + N_{A_1 A_2} + N_{A_2 A_2}} \tag{2.13}$$

where $N_{A_i A_j}$ is the number of $A_i A_j$ individuals, we can derive that the rate of change in p is

$$dp/dt = p(1 - p)(\tau_{\bar{A}_1} - \tau_{\bar{A}_2})$$

We can also derive that the change in average fitness with changing p is

$$d\bar{\tau}/dp = (\tau_{\bar{A}_1} - \tau_{\bar{A}_2}) + p[(d\tau_{A_1})/dp] + (1 - p)[(d\tau_{A_2})/dp]. \tag{2.14}$$

Since

$$(d\tau_{A_1})/dp = \tau_{A_1 A_1} - \tau_{A_1 A_2}$$

and

$$(d\tau_{A_2})/dp = \tau_{A_1 A_2} - \tau_{A_2 A_2}$$

then

$$(d\bar{\tau})/dp = 2(\tau_{\bar{A}_1} - \tau_{\bar{A}_2}) \tag{2.15}$$

Then, since

$$d\bar{\tau}/dt = (d\bar{\tau}/dp)(dp/dt) = 2p(1 - p)(\tau_{\bar{A}_1} - \tau_{\bar{A}_2})^2 \tag{2.16}$$

We can further note that the variance of $\tau_{\bar{A}_1}$ and $\tau_{\bar{A}_2}$ can be derived from

$$Var(\tau_{A_1}) = p\tau_{\bar{A}_1}^2 + (1 - p)\tau_{\bar{A}_2}^2 - \bar{\tau}^2$$

$$= p(1 - p)(\tau_{\bar{A}_1} - \tau_{\bar{A}_2})^2 \tag{2.17}$$

and hence the rate of change in fitness or value is merely twice the variance in average allele effects. This variance, as noted on p. 14, is not the total genetic variance but is equivalent to the additive genetic variance. The relationship of gain to this part of the genetic variance is part of Fisher's Fundamental Theorem of Natural Selection.

In the dynamics of allele frequency change, the dp/dt equation depends on $p(1 - p)$, which is positive except at $p = 0$ or 1, and on the sign and magnitude of $(\tau_{\bar{A}_1} - \tau_{\bar{A}_2})$, which is exactly equivalent to the numerator in the Δp equation. We can further note that $d\bar{\tau}/dp$ and dp/dt have the same sign for all values of p and are zero at the points at which $\tau_{\bar{A}_1} = \tau_{\bar{A}_2}$. Thus, if $dp/dt = 0$ at some intermediate value of p, $\bar{\tau}$ is maximized or minimized, and is a stable maximum if equilibrium p is stable or is an unstable minimum if the equilibrium p is unstable. Thus, not only does selection cause changes to proceed at a rate proportional to the genetic variance, but it maximizes the average fitness of populations. This completes Fisher's Fundamental Theorem of Natural Selection.

The inclusion of other variables into fitness, so that it is a function of such factors as population size rather than a constant, introduces many interesting and important features to the effects of selection. We review a few results here that are useful when considering selection under conditions of variable biotic or physical environmental elements and when considering variation patterns that natural selection may have generated.

One form of variable selection effect results from exogeneously caused spatial or temporal variations in the fitness of genotypes. These in general offer different possibilities for the maintenance of genetic variability and, when selection demes and mating demes are not identical, generate a wide variety of possible behaviors (Hedrick, 1987). Another form of variable selection effect that is less well analyzed results from variations in fitness driven by gene effects themselves on factors affecting fitness such as density or competition effects. Thus, to the allele frequency dynamics, we now add a population size dynamic and consider genotypic effects to be dependent on population size N. The system now expands to include a dN/dt and can be written as

$$dp/dt = \dot{p} = p(1 - p)(\tau_{\bar{A}_1} - \tau_{\bar{A}_2})$$
$$dN/dt = \dot{N} = \bar{\tau}N \qquad\qquad (2.18)$$

assuming that fitness is a function of allele frequency and population size, $\tau_{ij} = f(p, N)$ in general. Thus, the individual genotypic fitnesses or values may depend on population size, some having greater sensitivity to crowding than others, and may depend on which genotypes are doing the crowding. In most ecologically reasonable models, fitnesses above some minimal limit are generally negatively affected by density and hence $\partial\tau_{ij}/\partial N < 0$. Also, since the τ_{ij} may be functions of p, the $\dot{p}$ equation can be more completely written as

$$\dot{p} = [p(1 - p)/2][(\partial\bar{\tau}/\partial p) - (\overline{\partial\tau_{ij}/\partial p})]$$

where

$$[\overline{\partial\tau_{ij}/\partial p}] = p^2[\partial\tau_{A_1A_1}/\partial p] + 2p(1 - p)[\partial\tau_{A_1A_2}/\partial p] + (1 - p)^2[\partial\tau_{A_2A_2}/\partial p] \quad (2.19)$$

and any of the $[\partial\tau_{ij}/\partial p]$ may be non zero.

For purely density-dependent selection, $(\partial\tau_{ij})/\partial p = 0$, and the sign and critical points of p are determined by the sign and critical points of $\partial\bar{\tau}/\partial p$. Hence, for density dependence, $\bar{\tau}$ is maximized or minimized at $\dot{p} = 0$. For frequency dependence, however, this is not generally true. Thus, if specific competition, or resistance–virulence mechanisms, vary among genotypes depending on the genotypic frequencies of the other organisms, $\dot{p}$ may be zero at equilibria where population sizes or average fitnesses are not maximized (Selgrade and Namkoong, 1984; Namkoong and Selgrade, 1986; Roughgarden, 1979). In these cases, it is not uncommon for gene frequencies and population sizes to display quite intricate dynamics involving permanent fluctuations and even chaotic behavior.

The introduction of fitness functions to the discrete generation case and their difference equation models creates even more possibilities for unstable behavior (Asmussen, 1979). Further, the loss of random mating or equivalence of sexual productivity among monoecious genotypes introduces even greater instabilities in these models. Hence, in some open-pollinated seed orchard breeding programs, where sexual contributions are uncontrolled with neither equality among genotypes nor symmetry between sexes, the actual dynamics of gene and genotypic frequency change may be far more intricate than we imagine from our simple models.

2.5 Two-Locus Models

In the simplest case of independence of effects and frequencies, two-locus models are merely the summation of single-locus dynamics, fitnesses are maximized, and rates of change in fitnesses or other measures of value are proportional to the summed variances of average allelic effects. With epistasis, however, none of these results are realized, and although it is not developed for use in the breeding theories we later discuss, we briefly consider here how epistasis can affect our models.

We will consider a model that is a two-locus extension of the table on p. 17. We can write the fitnesses in values for the two locus genotypes as

	AB	Ab	aB	ab
AB	w_{11}^{11}	w_{11}^{10}	w_{11}^{01}	w_{11}^{00}
Ab	w_{10}^{11}	w_{10}^{10}	w_{10}^{01}	w_{10}^{00}
aB	w_{01}^{11}	w_{01}^{10}	w_{01}^{01}	w_{01}^{00}
ab	w_{00}^{11}	w_{00}^{10}	w_{00}^{01}	w_{00}^{00}

where the left element of both the superscript and subscript on the fitness w denotes the A allele content of each gamete in the zygote, and the right element denotes the B allele content. Then

$$w_{11} = w_{11}^{11} p_{11}^{11} + w_{11}^{10} p_{11}^{10} + w_{11}^{01} p_{11}^{01} + w_{11}^{00} p_{11}^{00}$$

$$w_{10} = w_{10}^{11} p_{10}^{11} + w_{10}^{10} p_{10}^{10} + w_{10}^{01} p_{10}^{01} + w_{10}^{00} p_{10}^{00}$$

$$w_{01} = w_{01}^{11} p_{01}^{11} + w_{01}^{10} p_{01}^{10} + w_{01}^{01} p_{01}^{01} + w_{01}^{00} p_{01}^{00}$$

$$w_{00} = w_{00}^{11} p_{00}^{11} + w_{00}^{10} p_{00}^{10} + w_{00}^{01} p_{00}^{01} + w_{00}^{00} p_{00}^{00} \tag{2.20}$$

assuming no position effects, and denoting the p_{jl}^{ik} for frequencies of $A_i A_j B_k B_l$ gametes, and $p_{jl} = \sum_{ik} (p_{jl}^{ik} + p_{ik}^{jl})$.

Then

$$\bar{w} = w_{11} p_{11} + w_{10} p_{10} + w_{01} p_{01} + w_{00} p_{00} \tag{2.21}$$

and assuming random mating, the gamete frequencies in the next generation p_{jl}' are

$$p'_{11} = [w_{11}^{11} p_{11} + w_{11}^{10} p_{10} + w_{11}^{01} p_{01} + w_{11}^{00} p_{00}] p_{11}/\bar{w} - r w_{11}^{00} [p_{11} p_{00} - p_{10} p_{01}]/\bar{w}$$

$$= [p_{11} w_{11} - r w_{11}^{00} D]/\bar{w}$$

$$p'_{10} = [p_{10} w_{10} + r w_{10}^{00} D]/\bar{w}$$

$$p'_{01} = [p_{01} w_{01} + r w_{01}^{00} D]/\bar{w}$$

$$p'_{00} = [p_{00} w_{00} + r w_{00}^{00} D]/\bar{w} \tag{2.22}$$

where D is the disequilibrium, and r is the recombination rate. It can be seen that all Δp_{ij} can be zero only if all of the average gamete effects are the same and equal to $\bar{w}$, and D must also be zero. For other values of $D \neq 0$, there could then be other equilibria that may be stable. Hence stable equilibria may exist where fitness is not maximized and where interlocus disequilibria are permanently maintained (Lewontin and Kojima, 1960). Investigations of various two-locus epistatic selection models indicate that up to seven equilibria may exist, two of which may be stable, and at the stable equilibria a marginal heterozygote superiority occurs (Karlin and Feldman, 1970) when taken as a one-locus average over all other loci. Even that weak condition may not be necessary (Ginzburg, 1983).

Thus, the dynamics of general epistatic two-locus systems are substantially more complex than those measurable by the one-locus margins. Further, the inclusion of any effects of nonrandom mating produces even greater complexity in the dynamics (Hedrick, 1980), in direct effects on genotypic frequencies as well as in indirect effects on recombination rates.

To advance to three and more loci, further complexities are added by the existence of three-way interactions and three-locus associations that are not merely the accumulations of all two-way effects. Thus, only great simplifications of interlocus effects would permit much detailed study of these higher dimensional dynamics. For example, Ginzburg (1983) assumed a strictly multiplicative effect between loci in order to assert that the average heterozygote fitness over all loci exceeds the average homozygote fitnesses. With a similar model, Franklin and Lewontin (1970) showed that multilocus disequilibria may be important in the evolution of whole sets of genetic loci. Even with these simplified models of interlocus effects, it should be remembered that we are dealing with that elusive element "fitness," and not with simple phenotypic measures.

2.6 *n*-Locus Models

In the face of such complexity of dynamical behavior, it seems obvious that much more theoretical work is needed to classify at least qualitatively different types of behavior or effect. At this time, however, it seems feasible to revert to the simplest definitions of gene effects and to use the one-locus model as an approximation. Thus, we assert that one-locus marginal effects in fact exist, and assume the approximation of independent effects among loci. For many traits, the approximation may not be poor, at least in the short run, but we must always be aware that the approximation is of unknown reliability.

We assume that because the genetic background for a gene action is variable and that the external environment varies among individuals, the marginal effect of each allele is obscured by an uncontrolled variance and that each gene has a small mean effect relative to that variance. For two loci of similar effect, we would expect a simple addition of the genetic variances and we would also observe that the average phenotypes for the two-locus genotypes occupy a wider range. For more loci, we simply accumulate marginal effect upon marginal effect and variance upon variance.

The effects of selection can then be modeled for each locus but with proportionately small effect at each locus. Thus, in the Δp equations (2.11) the s and t coefficients would be very small because the background variance is so large, and with a large number of genes, the genetic variance is so large that even directed selection will pick a favored genotype over another with large error and hence with small probability. A different way to express these effects is given by Griffing (1960), whose derivation of the change in value or "fitness" we follow.

For the value of a genotype with alleles i, j, we use the same concept and definitions as in Equation 2.7:

$$G_{ij} = \alpha_i + \alpha_j + \delta_{ij} \tag{2.23}$$

with the population mean

$$\mu = \sum_{i,j} p_i p_j G_{ij} \tag{2.24}$$

which we can scale without loss of generality to $\mu = 0$.

The average allele effects are

$$\alpha_i = \sum_j p_j G_{ij} \tag{2.25}$$

the total genetic variance is

$$\sigma_G^2 = \sum_{i,j} p_i p_j G_{ij}^2 \tag{2.26}$$

the additive genetic variance is

$$\sigma^2 \text{ (Additive)} = 2 \sum_i p_i \alpha_i^2 \tag{2.27}$$

and the dominance genetic variance is

$$\sigma^2 \text{ (Dom)} = \sum_{i \neq j} p_i p_j \delta_{ij}^2 \tag{2.28}$$

Also, identically as previously described for Equation 2.20, we determine the probability that an i, j genotype is selected to be a parent of the next generation. We can compute the means of the parental generation and the new allele frequencies from these probabilities. We then compute the progeny population value as determined for these new allele frequencies.

The probability of being selected depends on the genotypic value and on the relative amount of environmental variation for such quantitative effects, and on how stringently selection is applied. Consider that three genotypes with equal frequencies might be distributed along a measurement scale as follows:

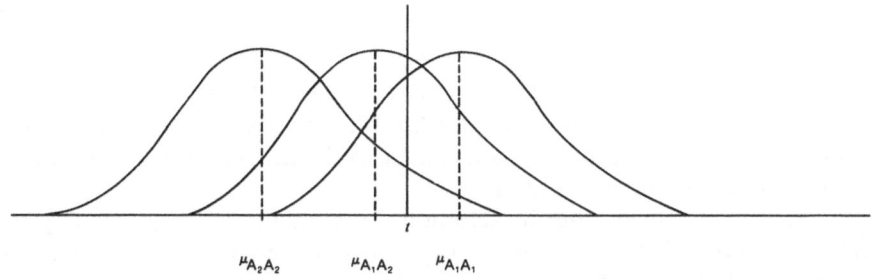

$$\mu_{A_2A_2} \qquad \mu_{A_1A_2} \qquad \mu_{A_1A_1}$$

If all genotypes below the truncation point of selection, t, are eliminated, the remaining genotypes would have a probability of being included that is approximately

$$Pr(\text{Select } G_{i,j}) = b[1 + (G_{ij}/\sigma^2)s]$$

where b is the selection proportion, σ^2 is the total genetic and environmental variance, G_{ij} may be positive or negative, and s is the difference in mean between the original population and the selected parental phenotypes. The relative frequencies of genotypes become

$$p_i p_j[1 + (G_{ij}/\sigma^2)s]$$

and hence, the average genotypic value of these parents is

$$\bar{G}_{\text{parents}} = \sum_{i,j} p_i p_j[1 + (G_{ij}/\sigma^2)s]G_{ij} \qquad (2.29)$$

$$= \sum_{i,j} p_i p_j G_{ij} + (s/\sigma^2) \sum_{i,j} p_i p_j G_{ij}^2 \qquad (2.30)$$

$$= 0 + (s\sigma_G^2)/\sigma^2 \qquad (2.31)$$

The expected genetic mean of the selected parents is the selection differential multiplied by a factor that is often called the "broad-sense" heritability.

Also, from the relative probabilities of selection, we derive the new allele frequencies:

$$p_i' = \sum_j p_i p_j[1 + (G_{ij}/\sigma^2)s]$$

$$= \sum_j p_i p_j[1 + ((\alpha_i + \alpha_j + \delta_{ij})/(\sigma^2))s]$$

$$= p_i + (s/\sigma^2)p_i\alpha_i \qquad (2.32)$$

Then assuming random genetic union, the new progeny population mean for two alleles would be

$$(p_i')^2 G_{ii} + 2p_i'(1 - p_i')G_{ij} + (1 - p_i')^2 G_{jj} \qquad (2.33)$$

substituting Equation 2.32 into Equation 2.33, we obtain:

$$\mu_{\text{progeny}} = \sum_{i,j} p_i p_j G_{ij} + (s/\sigma^2) \sum_{i,j} p_i p_j (\alpha_i + \alpha_j) G_{ij} + (s/\sigma^2)^2 \left[\sum_{i,j} p_i p_j (\alpha_i \alpha_j) G_{ij} \right] \quad (2.34)$$

$$\approx 0 + (s/\sigma^2) \sigma^2_{\text{Add}} \quad (2.35)$$

The approximate expected mean of the progeny generation is the selection differential multiplied by a factor called the "narrow-sense" heritability. Thus, in this conversion to a model that involves measuring phenotypes on some scale, we derive a similar relationship of rate of improvement to the genetic variance in that trait but modified for the change of scale.

Thus, for a single locus with its selection differential, s, and its possibly small heritability ratio, small changes in average allelic effects would be expected and thus also small changes in the genetic variances. In our simplified model with independent marginal gene effects and independent frequencies of occurrence, the selective effect is summed over loci and the totaled genetic variances also affect the rate of change in the phenotype. The total selection differential is then not the same for each of the n loci as for one, but is apportioned among them. It is then possible that large per-generation gains can be achieved if there are a large number of loci of small effect, without substantially changing the genetic variance, and with continued cumulative gains achievable for many generations (Mayo, 1980).

The successful experiences with long-term selection may result from at least three other mechanisms. A concept advanced for *Drosophila* and for maize (Dudley, 1977) selection experiments involves epistatic gene actions for both the existence of selection response plateaus and the existence of higher performance levels which can be achieved if linkages are broken and genetic variance is "released." With a multilocus model of sufficient complexity and with possible multiple alleles at some loci, the model involved would include the existence of several multilocus equilibria with some level of linkage disequilibria maintaining equilibria at some local maxima. While vague in specifics, such models are readily conceived of and are possible explanations for the patterns of progress for several generations: that is, plateaus in response during which more or less random mating is continued followed by a resumption of progressive response to selection, and so forth.

A third possible mechanism is that mutations accumulate and might be selected if their effects within these selected populations are favorable. Similarly, any changes in copy number of genes responsible for trait expression, or in the DNA or RNA control mechanisms that regulate transcription or translation rates, could simply create new genetic variation for advanced response capabilities (Hill and Rasbash, 1986a).

A fourth possibility is that no new genes are required but that loci not involved in initial response to selection may become involved within the nuclear and cellular milieu created during previous breeding cycles. These loci are epistatically related in that the effects of some loci impinge on a trait only after other loci have created a suitable genomic, cellular, tissue or organ-level environment. Such an epistasis might only be observed after sufficient changes in allelic frequencies occurred at other loci.

These four possible mechanisms may be difficult to discriminate in practice, and they may exist simultaneously in generating response to selection, but they do suggest different breeding practices. In the first and third concepts, continuous selection pressure induces small changes in frequency of each of many invariant loci or in the existence of more locus effects. In the second, occasional relaxed selection pressure is needed to permit recombinants to persist that may eventually allow the population to leave one local maximum for another. This might require the use of multiple initial populations to reach different local maxima. The fourth possibility would require continuous selection pressure but may also be benefited by the use of multiple initial populations to develop different multilocus optima. If the breeder must use a single population, it might be difficult to choose which long-term strategy to use. However, in all four cases, it would be necessary to use large population sizes in order to actively ensure the existence of alleles that generate the required variability. The uses of multiple populations, relaxed selection, and other techniques will be discussed in subsequent chapters on breeding methods.

2.7 Selection, Inbreeding: Migration, Mutation, Population Size

Thus far we have considered the effect of selection in the absence of any perturbing effects that the nonrandom association of alleles within and between loci might have or the effect of mutations on allele frequencies. Let us first consider the effects that limitations on population size have on allele frequencies and the effects of selection.

In any sample of individuals, the actual frequency of alleles obtained is rarely exactly equal to the expected or average frequency, and hence when there are only a few breeding individuals, there is a clear possibility of allele loss. With any history of selection, only consistent and strong selection for heterozygotes would keep alleles in the intermediate range. Otherwise, directional selection would drive alleles towards one extreme frequency or the other. Because the rate of change in allele frequencies is a function of $p(1 - p)$, the intermediate frequencies can be expected to be changing most rapidly, while the extremes would change more and more slowly as p approached 0 or 1. Thus, few loci would be expected to possess alleles in the middle range, while most allelic frequencies would be contained close to the extremes in a U-shaped distribution function. For N individuals, we can compute the probabilities that for two alleles, there are n_1', n_2', and n_3' individuals of the A_1A_1, A_1A_2, and A_2A_2 genotypes, and that these will then produce n_1, n_2, n_3 individuals in the next generation. For independent random individuals, this is a simple problem of computing combinatorials for multinomial distributions and for the probability of obtaining n_1', n_2', n_3' from n_1, n_2, n_3:

$$Pr(n_1', n_2', n_3') = \binom{N}{n_1 \, n_2 \, n_3} p^{n_1} 2p(1 - p)^{n_2}(1 - p)^{n_3} \tag{2.36}$$

Because p_1 is merely $(2n_1 + n_2)/2N$, the transition probabilities are constant, and we can conceive of simply building up all combinations $\binom{N}{n_1 \, n_2 \, n_3}$ of possible population states, giving rise to all other possible population states. Thus, we compute, say for just a few individuals, the probability of having no A_1 alleles, or 2 or 3, A_1 alleles, etc. We thus have a finite number of frequencies that can exist, and for each frequency, a finite number of other frequencies that can be reached. Thus, consider the probability of going from frequency i to frequency j as $Pr(i,j) = a_{ij}$. Then, for all possible states, we can form a matrix of transition probabilities:

$$[A] = [a_{ij}], \text{ with } 2N + 1 \text{ rows and columns.}$$

With such a matrix, and for any given array of individuals in the various states,

$$n = \begin{bmatrix} n_0 \\ n_1 \\ \vdots \\ n_{2N+1} \end{bmatrix}$$

we can compute the expected frequency array in the next generation:

$$n^1 = An^0 \tag{2.37}$$

or in say g generations:

$$n^g = A^g n^0 \tag{2.38}$$

Thus, with this view, we can compute various outcomes, and in particular can observe how limits on population size can ultimately drive populations to homozygosity even with random mating. A major question then is, how does this affect progress from selection? Obviously, with selection, the probability of populations starting at $p = 1/2$ and staying there is less likely, and one would expect the average gene frequency to have shifted to somewhat favor the retention of the favored allele.

For a model of additive gene effects:

$$\begin{array}{ccc} G_{A_1A_1} & G_{A_1A_2} & G_{A_2A_2} \\ 1 + s/2 & 1 & 1 - s/2 \end{array}$$

where

$$\frac{dp}{dt} = p(1 - p)(\mu_{A_1} - \mu_{A_2}) \tag{2.39}$$

$$= p(1 - p)s$$

and

$$Var(p) = \frac{p(1 - p)}{2N_e} \tag{2.40}$$

where N_e is the effective population size. Kimura (1962) derived the probability of ultimately fixing the favored allele (UPF):

$$UPF = \frac{\int_0^{p_0} \exp\left[-2\int_0^1 \frac{sp(1 - p)2N_e}{p(1 - p)} dp\right] dp}{\int_0^1 \exp\left[-2\int_0^1 \frac{sp(1 - p)2N_e}{p(1 - p)} dp\right] dp} \tag{2.41}$$

$$= \frac{1 - \exp[-4N_e s p_0]}{1 - \exp[-4N_e s]} \tag{2.42}$$

where p_0 = initial frequency of A_1 allele.

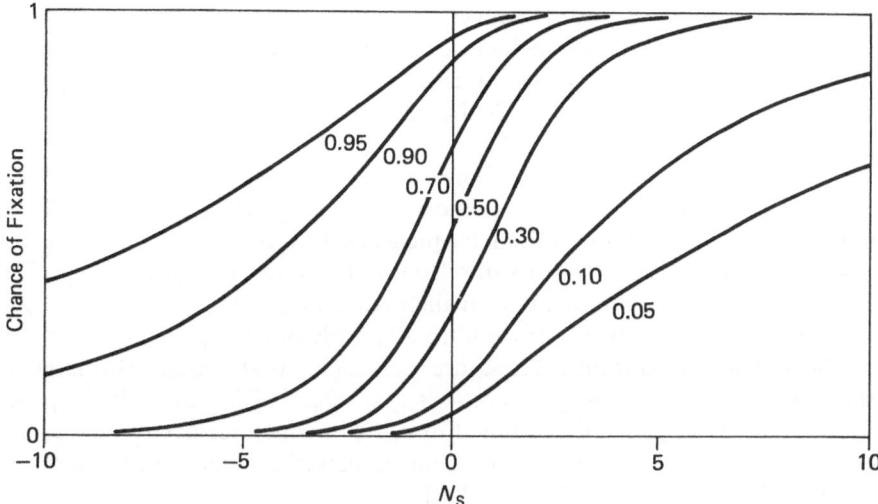

Fig. 2.1. The chance of fixation of a gene acting additively. The curves are often drawn for different initial gene frequencies (Robertson, 1960).

At $N_e s = 0$, UPF $\approx p_0 + 2N_e s p_0 (1 - p_0)$. However, at $N_e S \neq 0$, the UPF can be computed (Robertson, 1960) and graphed as in Fig. 2.1.

In this figure it can be readily observed that at $N_e s = 0$, the UPF of an allele is strictly proportional to its initial allele frequency, p_0. The s in this notation is equivalent to $(2s\alpha_i)/\sigma^2$ in Griffing's notation, and on a per-locus basis in a multi-locus model would be small. Thus, for $N_e s$ to be as large as 10, if s is less than 10^{-1}, N_e would have to be of the order of hundreds. With positive selection pressure of any magnitude, the chances of not losing a favored allele increases approximately proportional to $p_0(1 - p_0)N_e s$. Thus, for alleles in the middle frequencies, selection has its greatest effect, because $p(1 - p)$ maximizes at $p = 1/2$. For dominance models of intralocus effects, the slopes of the curves are somewhat different, and the response is proportional to $(2/3)N_e s(1 - q^2)$ if selection is for the recessive allele.

It can be seen that alleles at initially high-frequency present little problem in this maintenance; in fact, even with negative selection it may be hard to lose them and especially difficult to lose them quickly. However, at low frequencies, not only are the favored alleles easy to lose, but the effect of selection in decreasing their chances of extinction is small and hence large $N_e s$ is needed to save those alleles for use. It might be useful for breeders to consider that initial population sizes be kept large and that early generation selections endeavor to boost favorable but low allele frequencies so that allele frequencies move into an intermediate range. Once most alleles are into a safer frequency range, then selection could be relaxed or smaller populations may be affordable, but any severe population size reductions would have enduring effect on N_e and the probabilities of allele loss.

These models assume that quantitative gene loci are many, are of small individual consistent effect, and are the first of our models of genes affecting long-term response to selection. These models give little reason to expect variation in response among replicate populations. However, even for models of independent gene effects, nonuniform effects among loci can generate unequal responses to selection by starting at different initial allele frequencies and by having different magnitudes of effect. For alleles with initial $p = .5$, the effect of selection is strong, and although it is large at $N_e s = 5$, it is not much larger at $N_e s = 10$. Also, it is clear that the rate of allele frequency and phenotypic change depends on allele frequency, and that intermediate-frequency alleles provide most of the early generation gains. For loci with favorable alleles at low initial frequencies, low $N_e s$ levels do not greatly enhance their ultimate probabilities of selection. However, at higher $N_e s$ levels the response improves substantially. If such alleles then increase in frequency, they will be the ones to change most rapidly and to affect phenotypic gains most strongly, but only in later generations. These are the alleles that require the large $N_e s$ pressures in the early breeding generations and which, if increased in frequency, will provide the later generation gains. For cases in which recurrent mutation can generate genetic variation, N_e greater than 50 can allow continued response to selection (Hill and Rasbash, 1986a, 1986b).

We might alternatively consider that loci may display a range of s effects. For loci with large effects, at comparable initial allele frequencies a lower N_e is necessary to achieve a given UPF, and, especially at lower N_e levels, the larger effect loci are the ones that provide selection gain but also more variability among similarly treated populations. If there is a comparable potential gain for small-effect loci that are more numerous, they can be effective only if N_e is sufficiently high.

Because the precision of these predictions clearly depends on the quality of the approximations, at low N_e they can be inadequate. For $N_e > 8$, the errors in average gain predictions are not great (Hill, 1969). Obviously, linkages and epistasis can generate substantially different dynamics (Gill, 1965), but excluding those effects, gain predictions for $N_e > 30$ or $N_e > 40$ are not only reasonably accurate, but in practical breeding programs the loss of alleles can be low (Rawlings, 1970).

A series of studies by Frankham and others (1968a, 1968b, 1968c) is particularly enlightening. Those authors conducted replicated tests of a two-treatment factorial combination of population size (10, 20, and 40 pair matings per generation) and selection intensity (selection proportions of 10, 20, 40, 80, and 100%) over 50 generations. Over the first 12 generations, the strongest effect was that of selection intensity. Gain was clearly linearly related to selection percentage, except that the 80% selection level (four-fifths saved) was almost indistinguishable from the control population (100% saved). These results agree well with Kojima and Kelleher's (1963) observations. In contrast, the effect of population size is not as strong as that of selection level in the short run, but the larger population tends to attain a higher total gain at the same selection intensities. In fact, the 40% selection population with 40 pair matings did exceed the more intensive 20% selection carried with 10 pair matings. Also, in terms of the ratio of achieved gain to selection intensity (the realized heritability), the milder selections tended to exceed the more intensive percentages and also indicated that a longer period of response, and eventually a greater total response, may be obtained at the 40% selection level than at the 20% or 10%. It would thus appear that even in the short run, the increased gain by increasing selection intensity can be detrimentally affected if the populations maintained are small. While milder selection around 50% may provide a long, slow gain, at least the number of pair matings should be kept large.

In general, however, the response is a linear function of the initially estimated heritability and selection intensity and somewhat less of N_e. While the effects of selection intensity on realized heritability (i.e., achieved gain ÷ reach) are not clear, the early responses suggest that both small N_e and high selection intensities tend to reduce total gain. The variation among replicates of the populations was so high, however, that no one population could be expected to follow these average trends very closely. The smaller population sizes in particular exhibited great variations in response, indicating that at least sampling variations affect the replicate variance in the stochastic processes involved in the generational sequences. Even under these controlled environments, and with an organism adapted to those controls, the gene effect, selection, and mating processes generate substantial variations among identically treated (with respect to population size and selection intensity) population trials.

In the longer run, of 50 generations, most populations still appear to be responding to selection. The higher levels of selection intensity also still produce higher responses per generation. However, the effects of population size, which earlier were not clearly established, became a major factor in determining response. By the twentieth generation, there is a clearly established effect of larger population sizes on increasing response as lower selection intensities begin to exceed higher selection intensities if they also have larger population sizes. By the fiftieth generation, there is a rough equivalence in response of the main effects of increasing population size and of increasing selection intensity, and hence there is a much greater realized heritability for the larger populations. In addition, while all populations show some reduction in genetic variance, the larger populations continue to display a higher response rate than the smaller ones. It was clear that even with 16% heritabilities and using simple mass selection, the population responses of the larger populations under 10% selection exceeded the original mean by 1 standard deviation in two generations and by 2 standard deviations in five generations, with continued response after that. Thus, in relatively few generations, the population means far exceeded the original extremes.

In additional tests of Robertson's (1960) suggestions that early selection might advance gene frequencies into a safer intermediate range, smaller population replicates were split off from the 10%, 20%, and 40% selection populations with 40 pair matings each after 16 generations, by sampling 10 pair lines and breeding in those at their same selection intensities. All subpopulation splits of 10 pair lines immediately fell behind the larger population, and the lag accumulated. This is a clear experimental counter evidence to the concept that it is generally safe to restrict population size after an initial period of selection in a large population.

The variations among replicate populations, especially smaller populations, remained very large, tending to increase as a function of the mean response and hence increasing as selection intensity increased. The larger populations continued to exhibit less variation than the small ones. In addition, the variation among populations was exhibited when temporary plateaus and rapid responses alternated.

While the average response for the replicates at each of the selection intensities was reasonably smooth, individual replicates varied widely in size and period of response. The average declines in fitness, as tested in lines drawn from the selected lines and placed under relaxed selection, were moderate and lasted only a few generations. Therefore, there was only a moderate amount of natural selection opposing the directed selection. Some individual lines, however, did regress strongly because recessive lethals were still being carried and possibly also because of strong epistasis and linkages.

These long-term results indicate that epistatic interactions and the formation and destruction of linkage blocks can be important in holding genetic variations in populations, at times impeding, then aiding response to selection. The populations continue to respond to selection, and although there is some moderate decline in heritability, large variations among replicates also exist. In checking the state of the populations, it was found that some lethals were still present and affected selection response, but that the genetic variance was by no means exhausted. Several loci with large-effect alleles were still present at intermediate to high frequencies and some at low frequencies. Thus, it was concluded that several large-effect genes at low initial frequencies can continue to affect selection response long after one might otherwise assume their fixation. In addition, the presence of complicated linkage and epistatic effects can so confound the response to selection that useful genetic variance can also persist for many generations, especially at the larger population sizes and lower selection intensities.

For multiple-locus inheritance with independent loci of constant effect, the models for both short-term gain (Griffing, 1960) and long-term gains (Robertson, 1960) are reasonably consistent and provide some confidence that reasonably finite population sizes can effectively use most of the available genetic potential. This is not to say that these are the only kinds of gene actions that exist or are useful for breeding. Many kinds of epistasis would allow gains to be made, but because the effects are nonlinear, the directions of changes in gene frequencies are not constant. Thus, the selection pressure on alleles, even if additive within a locus, can change over the course of a few generations of breeding from positive to negative, while the phenotypic or fitness values continue to improve. If some alleles are neutral at the start of breeding but acquire selection effects either because of other gene effect changes or because ecological or economic conditions change, then population sizes would have to be carried that are appropriate when the initial $N_e s \approx 0$. For these cases, total population sizes required to ensure a low probability of loss of low-frequency alleles can be very large.

In Chapter 8 we consider in more detail sample sizes required to merely ensure the presence of alleles, but for now we assume that these numbers can be of the order of magnitude of the inverse of the frequency of the rarest alleles to be saved. These numbers are also affected by the number of alleles at that level, the mutation rate, and the number of loci carrying these rare alleles. Population sizes up to a few thousand may be required for these purposes. Once these alleles are saved, even at low frequency, their frequency can be increased by positive selection. Even if the direction of selection changes in the future, breeders can always select in different directions to increase the frequency of different alleles in different populations.

The worst case arises if a rare but favored allele is linked in repulsion to a disfavored allele in the population in which selection is intended to increase the favored allele frequency. In such cases, larger population sizes are required to hold the allele in the population until the linkage disequilibrium is dissipated.

Thus, the question of population sizes needed for breeding hinges on the distribution of allelic effects and frequencies. If all gene actions are the same and if natural selection and breeding objectives are consistent and the same, then there is no problem except maintenance of population structures. Thus, if natural selection has favored the heterozygote at many loci, and if breeding favors the same genotypes, or if directional selection favors the same alleles, little can be done to improve on natural populations except to simply maintain random individuals. To the extent that breeding and evolutionary objectives are different, however, or that some loci differ in gene action, then breeding can change performance levels for achieving directed breeding goals.

If there are many loci with alleles that have been under no selection, or under past selection regimes such that alleles with some beneficial marginal effect exist at low frequency, simple selection programs can be effective as long as $N_e s$ is kept large. Even with some forms of less independent gene action including epistasis and linkage, directed selection can be effective but may require some periods of relaxed selection while maintaining a large N_e. If single loci have large effect, or are initially at intermediate to high frequency, then their selection can be expected even while breeding for less effective and low-frequency alleles. If some loci are heterotic and stable at intermediate frequencies, then breeding selection will not change them, nor will they affect gain. However, inbreeding depression can exist that might be caused by epistatic interactions as well as by independent gene actions. If gene actions are independent, then any single-locus heterosis is likely to be at few loci and would be persistently counteracted by selection. If gene actions are independent but there are many loci with rare deleterious mutants (Bishir and Namkoong, 1987; Namkoong and Bishir, 1987), selection would purge the population of such alleles. If gene actions are independent but there is partial dominance at many loci, selection would still advance the frequency of favorable alleles although more slowly than with complete dominance. Therefore, inbreeding depression is not likely to be a substantial problem for selection unless epistatic effects and linkage are very strong.

2.8 Nonrandom Mating and Selection

Just as inbreeding affects the genetic variation in populations and as linkage affects two-locus epistatic dynamics, so nonrandom mating substantially affects general selection models. By nonrandom mating we mean any effect that results in genotypes having frequencies that are not those expected to be generated by independent association of alleles. Such effects may be caused by nonrandom mating structures such as sib mating, selfing, or any form of genotypic assortative or disassortative mating as the result of coincidental flowering, reproductive organ selection effects, or merely presence in a common mating deme. It may also be caused by the sexuality of individuals that generate disproportionate gametic contributions in the next generation. For monoecious individuals, this might simply mean a more fertile female or male; for dioecious individuals, however, not only disproportionate total gamete production, but also sexual asymmetry and differential selfing can result. In mixed gender populations, the effective gamete production depends on the frequencies of all sexual types and, hence, strong effects on changes in genotypic frequency independent of viability selection and of any propensities to mate with relatives or within mating demes (Müller-Stark, 1985).

These effects are to be distinguished from the stochastic variations in gametic frequencies caused by finite population sizes. The finiteness of populations can cause drift in allele and gamete frequencies and hence variation among populations that are not caused by any association with selection. Thus, average allele frequencies are expected to remain the same over many populations that either do exist or can be expected to exist, but any one or few populations may have 'drifted' to different frequencies. While random association of alleles within demes may exist, if p_i differs among the i populations, the homozygote frequency within populations, p_i^2, when averaged over n populations, $\sum p_i^2/n$, is greater than the square of the overall average allele frequency, $(\bar{p})^2$. This can be seen by examining the variance of p_i over n populations:

$$Var(p_i) = \frac{\sum p_i^2}{n} - \bar{p}^2 \tag{2.43}$$

Since $Var(p_i) > 0$, the average homozygote frequency is greater than the expected frequency of homozygotes, $\bar{p}^2$, which would be expected if population subdivisions did not exist. Known as the Wahlund effect, this creates excessive homozygotes even with 'random' mating within populations. In this sense, there is a correlation of allelic occurrences within loci that can be related to an 'inbreeding' coefficient but does not involve preferential mating within demes.

Two-locus associations are far more complicated, but the various kinds of association and their effects on variances and covariances are definable (Cockerham and Weir, 1973). Here an additional problem is introduced by the failure of any initial linkage disequilibrium to be immediately dissipated by random mating within populations. Although an average disequilibrium would be expected to be zero over the complete set of populations, any one population would carry allelic contributions among loci that reflect both past and present sampling deviations from independence; hence, two-locus genotypic frequencies with different levels of correlations would exist.

The model of independent loci, however, is often still useful as a paradigm and is often used for estimating effects of selection and probabilities of allele loss. For this model with stochastic sampling effects, a positive F coefficient is a reflection of excessive homozygotes from finite gametic sampling. If all genotypes contribute an equal number of gametes to the next generation, from N monoecious parents, there is a $1/2N$ probability that random genetic union would result in a zygote with gametes from the same parent, and hence decrease heterozygote frequency. This results in a change in the F coefficient such that

$$(1 - F_t) = (1 - 1/2N)(1 - F_{t-1}) \qquad (2.44)$$

In an ideal random mating population, there is a steady progression in F, and as long as N is finite, heterozygotes are lost; so long as that is so, there is always a probability that an allele will be lost, and F inevitably approaches 1.

For some populations, the increase in F may be more rapid than expected by Equation 2.44 because of undirected sampling effects such as are caused by unequal family size of the N parents. Then, to express the rate of change in F, a number different from the census number of parents, N, can be substituted. Consider for example that seed yield may be independent of any gene of interest, yet varies randomly among trees. If the population replaces itself, the next generation will, on average, have more homozygotes than suggested by Equation 2.44. Thus, if the average family size produced is $\bar{k}$, then N parent trees could produce $N\bar{k}$ trees and $N\bar{k}(N\bar{k} - 1)/2$ pairs of mates. Of all possible such pairing, each family would produce $k_i(k_i - 1)/2$ pairs in which both parents came from the same family. In an idealized population, these pairings within families would occur only $1/N$ times, and by so defining an effective N_e:

$$1/N_e = \frac{\sum k_i(k_i - 1)}{N\bar{k}(N\bar{k} - 1)} \qquad (2.45)$$

Since $\sum k_i^2 - \bar{k}^2 = \sigma_k^2$, we can rewrite this equation as

$$N_e = \frac{N\bar{k}(N\bar{k} - 1)}{(N - 1)\sigma_k^2 + N\bar{k}^2 - N\bar{k}} \qquad (2.46)$$

If $\sigma_k^2 = \bar{k}$, as would be the case if k were Poisson distributed, then $N_e = N$. If k_i was controlled so that $\sigma_k^2 = 0$, then $N_e = 2N$.

Another way in which the census number would not accurately reflect rates of loss of heterozygotes under conditions of restricted parental matings is when unequal numbers of dioecious males and females exist. In such cases:

$$N_e = \frac{4N_m N_f}{N_m + N_f} \qquad (2.47)$$

If $N_m = N_f = N/2$, then $N_e = N$. Other effects, such as overlapping generations or differential reproductive rates among age classes would also decrease N_e relative to N (Giesel, 1971). This, N_e is called an effective population size and, specifically, the "inbreeding effective number" because of its relationship to the progress of inbreeding and its focus on the parental population.

Another similarly motivated measure of effective numbers is also an abstraction related to the census number that would exist in an idealized situation for the variance in gene frequency to be as expected. The concern in this case is that the progeny population is expected to display variations among their subpopulations according to a binomial distribution, $\sigma_p^2 = \bar{p}(1 - \bar{p})/2N$ for subpopulations of size N. However, the same kinds of influences that make N a biased number for predicting inbreeding result in N being not a good number to use to predict σ_p^2. For the case in which family sizes may vary, the N_e required to satisfy

$\sigma_p^2 = \bar{p}(1 - \bar{p})/2N_e$ is:

$$N_e = \frac{2N}{\sigma_k^2(k + F)/\bar{k} + (k - F)} \qquad (2.48)$$

Again, F is the departure of zygotic frequencies from Hardy–Weinberg equilibrium frequencies (Crow and Kimura, 1970).

If populations undergo sequential, temporal variations in N, then the variance of p will also change over generations. In order to determine the N_e that would give an average estimate of how σ_p^2 was generated for n such generations, we can estimate the sampling variance for the sequence as

$$\sigma_p^2 = \frac{\bar{p}(1 - \bar{p})}{n}\left(\frac{1}{2N_1} + \frac{1}{2N_2} + \cdots + \frac{1}{2N_n}\right) \qquad (2.49)$$

and define N_e as satisfying $\sigma_p^2 = \bar{p}(1 - \bar{p})/2N_e$. Then equating the two σ_p^2 estimates yields $N_e = [n/\sum_i(1/N_i)]$, the harmonic mean of the various N_i. The harmonic mean, being more strongly affected by low numbers than is the arithmetic mean, gives lower values than the arithmetic mean, and therefore indicates that the effective population size is strongly affected by bottlenecks of low N_i. The two types of N_e are expected to converge as long as the random nature of unequal gametic contributions is maintained. For closed tree populations with mixed age classes and repeated matings by individuals that may mate across generations, Hill (1972) further derives a variance effective number for these closed populations. Emigh and Pollak (1979) consider different life tables for the two sexes, and derive a different estimate for the effective population size.

If genotypic differences in fertility or fecundity exist, then those genes, or genes that are in linkage disequilibrium with them, would have substantially different dynamics. In these cases, we could not simply consider that the dynamics of independent, quantitative loci can merely be adjusted for a smaller N_e nor that progress to ultimate fixations or stabilities can be reasonably well approximated. In fact, stable equilibria can be destabilized or disappear entirely, and new equilibria created solely by the influence of sexual asymmetry. Such asymmetries are very likely to be a regular feature of natural forest reproduction and even of breeding populations generated by seed orchards (Müller-Stark, 1985). The effect is caused by genetically related differences in male and female reproductive efforts, responses of reproductive organs to environmental variations, timing of both male and female gamete flight and receptivity, and so forth. Even in seed orchards, it is not uncommon for certain clones to predominate as males and for certain specific clones to dominate seed production. Thus, without strong control of reproduction in even well-cared-for and designed orchards by controlled pollination or supplemental pollination schemes, substantial departures from expected gain may result (Friedman and Adams, 1985).

Finally, we consider that at least some forest populations may be primarily composed of partially isolated populations, that evolved for a long time with different average allele frequencies. Even in the absence of direct selection effects, genotypic and allele frequency changes can occur. If we consider random mutation, for example, with a mutation rate μ, for changing from an allele A_1 to A_2 and a rate, δ, for changing A_2 to A_1, then a mutation equilibrium p would be expected at

$$p = \mu - \delta \tag{2.50}$$

With the constraints of the stochastic effect of finite population size, however, various outcomes are possible, and the probability density $Pr(p)$, that a particular p is found, is proportional to

$$Pr(p) = 2N_e(1 - p)^{(4N_e\mu - 1)}p^{(4N_e\delta - 1)} \tag{2.51}$$

These same equations can be derived for migration rates that effectively substitute one allele for another.

One peculiarity of this form of the equation is that if $4N_e\mu = 1 = 4N_e\delta$, or $\mu = \delta = 1/4N_e$, then $Pr(p)$ is proportional to $2N_e$ and is no longer a function of p, and therefore $Pr(p)$ is uniform for all p and all gene frequencies are equally likely. Therefore, such mutation or migration rates are sufficient to hold almost all allelic frequencies equally likely, and therefore these rates can maintain polymorphisms in spite of tendencies to drift to fixation.

On the other hand, if N or μ is large such that $4N_e\mu > 1$, then

$$Pr(p) \approx 2N_e(1 - p)^{f_1(N_e, \mu)}p^{f_2(N_e, \delta)} \tag{2.52}$$

is a function of $p(1 - p)$ that has a peak in the intermediate values of p. In fact, at very high values of N the solution for $Pr(p, t)$ is proportional to

$$Pr(p) = \mu[\ln(4N_e\mu - 1)]p^{4N_e\delta - 1}(1 - p)^{4N_e\mu - 1}$$
$$+ \delta[\ln(4N_e\delta - 1]p^{4N_e\delta - 1}(1 - p)^{4N_e\mu - 1} \tag{2.53}$$

which is close to zero in every case except at $p = \dfrac{\mu}{\mu + \delta}$, which was the deterministic solution we previously reached.

We can also see that if $N\mu$ is very small, the solution for $Pr(p) \approx (1 - p)^{-1}p^{-1}$, is the reciprocal of the peaked quadratic function which has a deep concavity in the intermediate ranges of p. Hence, if $N\mu$ is small, the random processes of drift lead to fixation of the loci for one or the other allele. If N_e is low, drift occurs without much influence of otherwise effective mutation, or of migration.

The effects that limited population sizes jointly exercise on selection and mutation/migration can also be examined. The resultant probability distribution for P is

$$Pr(p) \approx 2N_e e^{(-4N_e sp)}p^{(4N_e\delta - 1)}(1 - p)^{(4N_e\mu - 1)} \tag{2.54}$$

This function now shows that selection and mutation/migration independently have simple product-like effects on the probability distribution, such that a large s can push the distribution to the right-hand state as long as N is large enough and the effects of mutation or migration on increase of the alternate allele are not high. Consider, for example, that N_e is large but that both μ and δ are small so that $4N_e\delta = 4N_e\delta \approx 1$. $Pr(p) \approx 2N_e \sim \exp(-4N_e sp)$, which is an increasing function of p and hence tends to decrease the probability of having low gene frequencies and increases the probability of high frequencies. It also indicates that, if s is of the order of $1/4N_e$, even at low initial p, the population can keep the favored allele.

On the other hand, if $4N_e s$ is very low, such as at $e^{-4N_e s p} \approx 1$, then $Pr(p)$ can be largely determined by the balance between drift and the effects of mutation/migration. Then, the introduction of new alleles through mutation/migration, even at a low frequency, will maintain genetic variability. Thus, if migration may exist, very strong isolating mechanisms are required if populations are to diverge in the kinds of alleles carried. On the other hand, alleles can be lost under these conditions if breeding populations are developed with either low N_e or low s.

The effects of migration, finite population sizes, and selection, even for the single-locus models, can thus be seen to have intertwined effects and could affect the potential advances expected of breeding efforts. Unfortunately, these are the very kinds of effects that are present when pollen or seed contamination of breeding populations is allowed, especially when the effective number of parents is low.

CHAPTER 3

Basic Concepts in Recurrent Selection

In Chapter 1, we briefly discussed the difference between agricultural crop breeding and forest tree breeding. We presented the need to broaden the concept of tree breeding to include the ability to handle a multiplicity of objectives and environments. We also discussed mechanisms of natural selection, and concepts of gene expression and environmental variation in selective breeding, to emphasize the need to breed trees within an economically and ecologically sound context. In Chapter 2 we introduced basic population genetic concepts and emphasized the importance of knowing the dynamics of population-level events to enhance our capability for making sound breeding decisions.

We thus view tree breeding as a means of coping with different unknown or variable circumstances arising in forest management. In our view genetically improved planting stocks are not the ultimate goal of tree breeding, but only represent the transient products that are useful for coping with particular circumstances. The goal of tree breeding is to provide means for balancing different forestry goals, while using available techniques, and to generate multiple products by developing genotypes useful for those ends. However, the value of forests and of the effects of different breeding procedures varies, depending on such factors as product demand, species abundance, organizational interest, ecological constraints, etc. In the following section we discuss some objectives of tree breeding that influence the value system. In subsequent sections, we discuss a basic model for recurrent selection, gains from multiple-generation breeding, limits to selection, some problems in developing recurrent selection programs, and, finally, the uses of vegetative propagation.

3.1 Changing Roles of Tree Breeding and Forest Genetics

Humans have long used heritable variations in domesticated species. It might reasonably be conjectured that the domestication of animals and plants was largely informal and accidental at first, but that with the advent of stable agricultural systems perhaps 5,000 years ago more deliberate selection of good performers became more common. Even today, sustained improvement of varieties and surveys for potentially useful species and varieties continues, using methods at levels of sophistication that may vary from those used several thousand years ago to those designed to utilize different gene actions, for several traits, in a range of environments.

The initiation of tree breeding has been more deliberate and formal. As discussed in Chapter 1, tree breeders adopted intensive breeding as an extension of agricultural breeding methods with minor adjustments. The dominant driving force in tree breeding has been short-term economic gain. Such economic justification has been critical for establishing tree breeding as an acceptable part of forest management, but the principle is too limited to be useful in coping with the multiplicity of problems in forestry affected by genetics. Nevertheless, the principle of maximizing short-term economic gain has affected all aspects of tree breeding. This principle has narrowed the perceived scope of responsibility of tree breeders and forest geneticists and the types of breeding techniques they use, and has influenced the way they manage breeding populations. Although there are many precedents for carrying out breeding for reasons other than short-term economic gain, the merits of other objectives for breeding have been obscured (Box 3.1). This principle also leads to the perception that multiple-generation breeding is a simple repetition of short-term breeding activities, and so long as breeders continue to maximize short-term economic gain, multiple-generation breeding is no different than merely iterating short-term programs (Kang, 1982).

Box 3.1 Breeding Forest Trees in Wisconsin

Of the seedlings produced in Wisconsin nurseries in 1987, 62% were red pine, 13% were white pine, 4.6% were jack pine, and 20.4% encompassed 12 other species (Wisconsin DNR, 1987). Despite relatively little genetic variance in red pine, gains of 3–4% in height growth and 9–11% in stem volume are suggested as possible with family selection of the top 10% (Ager et al., 1982). If we disregard white pine breeding for growth (because of white pine blister rust), then the genetically diverse jack pine (Rudolph and Yeatman, 1982), will be the only other *Pinus* species planted in Wisconsin. However, assuming that current planting trends continue, it is not difficult to conclude that for short-term economic gain a 3–4% gain in red pine would be more beneficial than a 20% gain in jack pine. If only short-term economic gain mattered, then red pine would be the only species that would be worth breeding. However, tree breeders who are familiar with the capacity of jack pine to respond to advanced breeding techniques insist on breeding jack pine because its long-term potential can be technologically improved. Other species may also be developed by low cost breeding methods for inclusion in planting programs, while some species can justify intensive programs for short-term gain immediately.

In fact, the Wisconsin Department of Natural Resources (DNR) is breeding jack pine as well as red pine. Justification for jack pine breeding hinges on technically facilitating the expectation that progress from multiple-generation breeding will help change the future planted species distribution in a manner that is both economically and ecologically satisfactory. Fulfilling this expectation, however, will require multiple-generation selective breeding, and may involve techniques such as the development of inbred lines (Rudolph, 1981) that are not seriously considered for other Northern conifers.

However, as advanced generation breeding populations have developed around the world, it has become clear that a separate view of long-term breeding is needed that is not a simple extension of short-term economic gain concepts. Further, increasing concern for the environment has begun to challenge foresters and tree breeders to reevaluate their responsibility and role in managing forests within the context of ecosystem management (see Chapter 1, Sections 1.3.2 and 1.4). In recent years, this change in perspective for tree breeding appears to have gained momentum, and implies a need to reexamine the role of forest genetics and tree breeding in this changing world.

The role of forest genetics and tree breeding may be divided into two broad, traditional forestry categories: natural regeneration and artificial regeneration. In the past, little emphasis was given to the need to consider genetic consequences of natural regeneration. As discussed earlier (Section 1.4), natural evolutionary processes cannot be as sharply differentiated from tree breeding as they can be in domesticated plant and animal breeding. Therefore, lacking knowledge of evolution and of the natural ecosystem, breeders would develop concepts of how to guide evolution without benefit of available information and concepts. Conversely, breeders well informed about natural selection and evolutionary processes can and should contribute to management of naturally regenerated populations and explain potential courses of evolution on managed and unmanaged forest lands. By such management practices as controlling the sizes, numbers, and temporal sequences of reproductive patches, the distribution of genetic variation can be controlled. By supplemental seeding or planting with propagules selected for specific traits or for the diversity they can introduce, more direct influence can be exercised. In all subsequent discussions, we assume that selection goals for these less intensive systems, and for ecosystem productivity, are as important as are goals for fiber or timber production.

In the past, silviculturists have applied cultural practices to natural stands and made decisions about harvesting trees without considering genetic effects. By so doing, they selected trees and thus influenced the mating structure of populations. In many cases, silviculturists have thus practiced breeding without the benefit of input from geneticists. To practice genetically sound management of natural stands, extensive knowledge of population genetics and evolution of natural populations is useful. In some areas, such as in the National Forests in the northeastern United States where natural regeneration is the primary means of reproducing forests, there is a serious need for geneticists to work with silviculturists in managing the land (Murphy and Kang, 1985). In other areas, such as some national forests of the western United States, where natural regeneration is used on as much as 80% of the Ponderosa pine types, geneticists and silviculturists work together in applying natural regeneration prescriptions (Friedman, 1986).

Even for naturally regenerated species, maintaining breeding populations apart from forest stands is desirable to augment or temporarily substitute (i.e., single-generation planting) in natural regeneration areas as the need arises. To create and manage such breeding populations, breeders must have breeding strategies that differ from those applied to species deployed for intensive breeding.

Breeding for artificial regeneration may be further subdivided into long-term breeding and short-term activities (Fig. 3.1). Although long-term breeding necessarily requires multiple-generation breeding as its primary mechanism, long-term

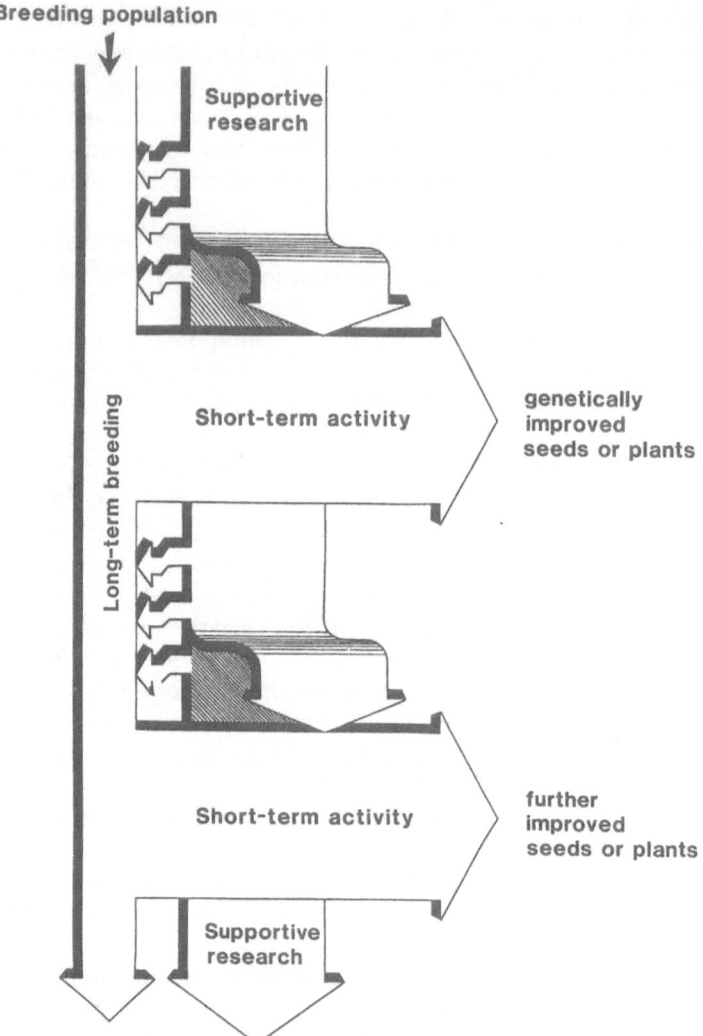

Fig. 3.1. Functional components of a tree breeding system. (From Gullberg and Kang,1985a.) Long-term breeding is aimed at maintaining an optimum balance between continuous genetic improvement and diversity conservation by manipulating the long-term breeding population in a structured manner. Short-term activity is further divided into short-term breeding and multiplication. Short-term breeding manipulates long-term breeding populations through intensive testing, selection, and mating efforts to maximize possible genetic gain with respect to economically important traits of the time. The products of such short-term efforts are subsequently multiplied. Research supports both long- and short-term breeding. Long-term supportive research tends to emphasize solving inherent problems in tree breeding. Short-term supportive research tries to develop means of maximizing short-term economic gain.

breeding can be differentiated from short-term activities in some other respects (Gullberg and Kang, 1985a). The most notable distinguishing features are (Kang and Nienstaedt, 1987): (1) many long-term breeding activities cannot be justified on the basis of short-term economic merits; (2) the primary goal of long-term breeding is to make future short-term breeding continuously successful under changing environments and societal demands; (3) long-term activities tend to focus on inherent structural problems of tree breeding, such as the diversity of breeding populations, loss of useful alleles, the generation turnover period, and the purging of deleterious alleles, whereas short-term breeding focuses on gain, inbreeding depression, and immediate objectives; (4) long-term breeding does not necessarily imply activities that are continuous over many years, because some will require less time than a short-term breeding cycle and because many activities can be completed at a juvenile stage of tree growth; and (5) population management concepts and technical recommendations for long-term breeding will frequently contradict those of the short term, which often implies a repetition of single-generation techniques (Kang, 1982). Examples of long-term breeding include jack pine breeding (see Box 3.1) and maintaining breeding populations for naturally regenerated species as well as for species having intensive breeding programs with separate long-term breeding components (Barnes, 1986; Gullberg and Kang, 1985b).

Short-term activities (as defined in Fig. 3.1) usually depend on one or two additional cycles of breeding, and trees produced in the short term usually do not feed back into the long-term breeding populations. Short-term activities also include multiplication efforts such as seed orchards, clone orchards, and tissue culture. As is discussed in Section 3.5, most vegetative propagation techniques are used in short-term activities, and are not breeding.

Of the various kinds of activities discussed, long-term breeding is the area of main interest in this volume. Because of the connection of long-term breeding with natural regeneration and short-term activities, we discuss those activities as needed. However, in discussing breeding techniques, plans, and strategies in subsequent chapters, we assume multiple-generation breeding unless otherwise specified. We also assume that the paradigm organization has opted for long-term breeding and must operate under certain economic limitations. Therefore, when we discuss different alternatives and assign costs we are referring to the organizational limitations, and do not discuss the reasons such limitations may exist.

3.2 Basic Model and Some Simple Models of Recurrent Selection

We assume that the array of species we are to be working with has already been chosen and that the appropriate source populations are known. Because these assumptions are not true for many forest management systems, we will discuss the deviations from these assumptions further in Section 3.3 and in subsequent chapters.

A complete cycle of selection, which is the basic unit of any recurrent selection system, is shown in Fig. 3.2. The basic cycle is composed of two populations, base and selected, and two activities, selection and mating. Breeders vary the techniques,

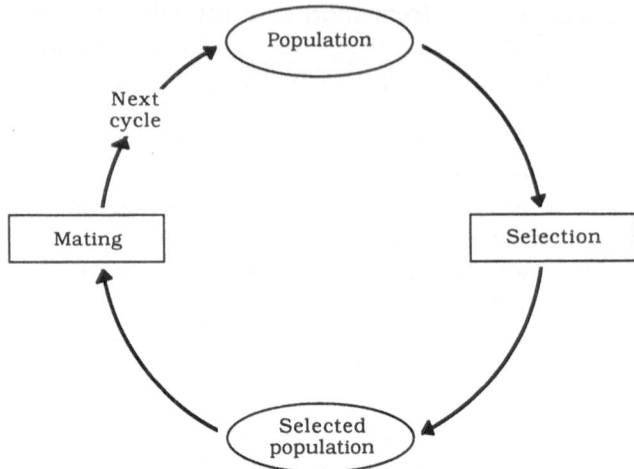

Fig. 3.2. Basic model of a complete cycle of selection.

manipulate the number of populations involved in the cycle, and may treat the number of concurrent cycles differently. Accessory populations and special techniques may be attached to a cycle without necessarily becoming integral parts of a breeding cycle. For example, if progeny testing is used to evaluate parents to be selected, but the test offspring are not involved in the subsequent mating, then the progeny test population is not an integral part of the breeding cycle (Fig. 3.3). Repeating a particular cycle over generations generates a recurrent selection system. The breeder's goal, then, is to determine the type and organization of such cycles that would generate the best results with respect to the parameters of interest.

We begin our consideration with the basic breeding system, simple mass selection in which seed is collected from the best individuals as judged by their field performance. For example, when collecting seed for storage for subsequent planting of an annual crop, the best yielding individuals are used as the seed source, while the rest of the crop may be eaten or otherwise used without the seed being saved. Direct observation of the trait of interest is implied, and no control of pollen source is assumed. In forest tree breeding, this would typically involve merely observing or measuring the best harvest-aged trees and collecting the open pollinated seed for planting in nurseries or for direct seeding for forest regeneration (Fig.3.3a). Such systems are relatively inexpensive but require the breeder to wait for the trees to naturally achieve sufficient seed productivity or reach an age sufficient to judge performance, whichever is later, before the next generation can begin. Nevertheless, such breeding can succeed in producing cumulative gain and has been suggested as a potential means of breeding white spruce in the Lake States (Nienstaedt and Kang, 1983). So long as the effective population size remains large, then additive genetic variation can be used and the population can iteratively improve. However, because the pollen contribution to the seed is usually assumed to be a random and wide sample from the general population, the gain is achieved only on the female side. For monoecious species with equal sexual expression, the effect is merely to halve the gain achievable by selecting both parents. We consider these forms of

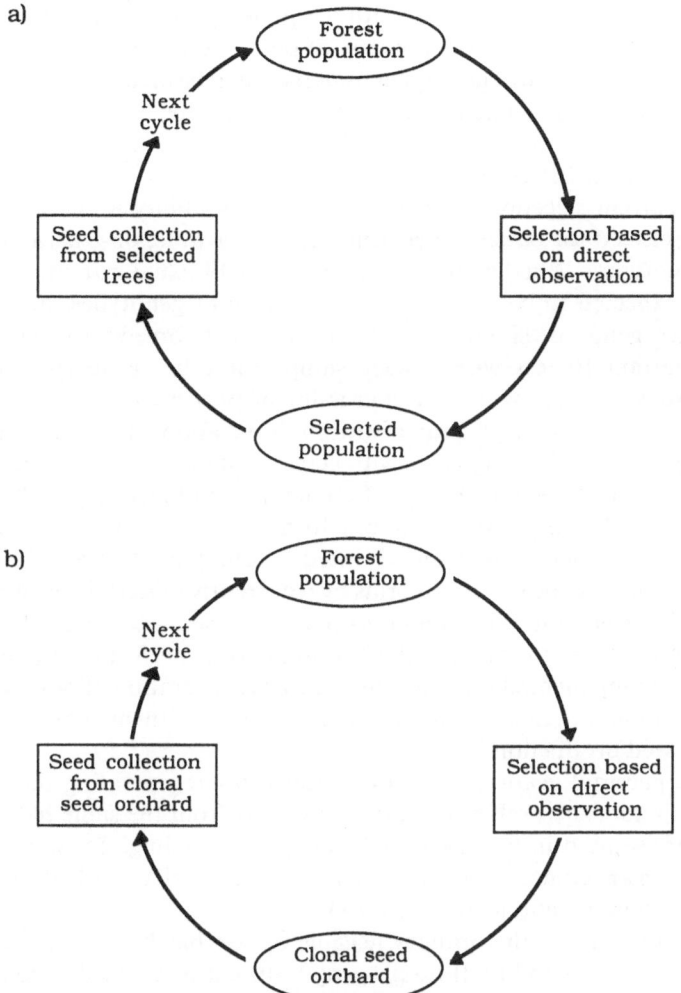

Fig. 3.3. Simplest selection cycles. (**a**) The forest population at the bottom of the cycle may be a rogued (or partially harvested) forest plantation at the top of the figure. (**b**) Clonal seed orchard consists of clones settled in the forest population.

breeding to be a basic model, only slightly more involved than seed tree selection methods from some local or other good provenance source. In fact, regeneration costs may be comparable for these simplest forms.

When the control of pollination is feasible, it is possible to select both parents. This simple recurrent selection (SRS) involves the same kind of iterative procedures for cumulative improvement of the breeding population as does mass selection, but pollination control generally offers other opportunities for testing and estimating gene effects. The development of the breeding population from only the selected parents of the previous generation and iterating the selection procedure on the subsequent progeny is the same but with the pollen parents also selected. Because observations can be made on the progeny of potential parents, some retrospective testing of the performance of genotypes as parents can be made in addition to simple observations of their own phenotype. Thus, the SRS breeding populations can be augmented in value by more precise selection of the breeding value of parents, as well as by biparental selection.

Simple recurrent selection for general combining ability implies that the individuals are selected for parental performance with a wide, or general, set of other parental genotypes. Whether female, male, or a bisexual parent, the alternate parents are expected to be a nonspecific collection of genotypes, or, at least, the other parent's genotype should not be relevant to its breeding value. Thus, it is usually important to test with a wide sample of other genotypes by multiple crossings. However, even without test crossing or progeny tests, it is assumed that the next generation will be generated with multiple and widely dispersed parental intercrossing. Thus, SRS progresses with general intercrossing of selected parents.

Simple recurrent selection for specific combining ability is possible for those species in which breeding has advanced to a level such that an established or standard line or variety is consistently used as one parent in a hybrid breeding program. To improve the yield of hybrids by recurrently or iteratively improving the nonstandardized parent(s), the standard parent is used as a tester. The potential parents are selected on the basis of the hybrid progeny performance, and these are intercrossed among themselves to produce the next generation of potential parents. In any generation, the selected parents, or some subset of them, may be used for the commercial seed production.

When the parental genotype cannot be saved for regenerating the next generation's population, their selfed seed, or saved seed from the same half- or full-sib families as those used in testing, may be used for breeding. These forms of sib–family selection are common in annual plants and are well described elsewhere (see for example, Hallauer and Miranda, 1981).

For most forest trees, the primary feasible option has been some form of SRS for general combining ability. If no progeny testing is performed, a common form of breeding has been to select individuals ("plus trees") on the basis of their own performance, and to intermate only among these selected parents for commercial seed production or for creating the next generation of breeding populations or both. This sometimes involves the creation of a clonal seed orchard (Fig. 3.3b). The cost of this biparental SRS may not be much greater than for the most basic model, yet may offer much greater gains and ancestral control. However, heavy pollen contamination of seed orchards can reduce gains to mass selection levels (Friedman, 1987).

If a progeny testing phase is imposed for testing either the general or specific combining abilities, some subset of the forest population must be prescreened and test crossed and the test progeny observed before the parents are selected for intermating: Then, the rest of the breeding operation can continue (Fig. 3.4). This phase requires a major increase in capital and temporal costs and represents a major operational difference between low cost and intensive breeding programs.

Several operational variations on this basic pattern may be used depending on

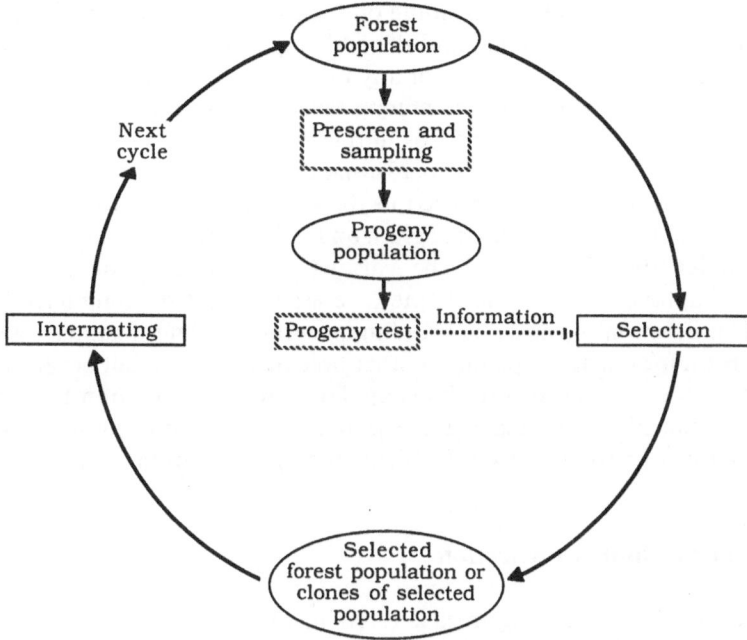

Fig. 3.4. Simple cycle with progeny testing.

the cost and time required for screening and testing and on the size of the genetic variances. For most tree breeding, the efficiency of recurrent selection depends on the additive genetic variance, and, hence, the necessity for screening and testing depends on how well additive effects or additive breeding values can be observed; if dominance gene effects are large, even inherited differences may not be transferred to progeny. If the breeding values are difficult to observe because environmental variation substantially masks genotypic differences, then prescreening may be ineffective and progeny testing may provide the only means of selecting superior parents. This may occur not only because of variation in the external environment, but also because the managed environments that are anticipated for the improved forest populations are different or individuals at the age of expression of important traits are not observable.

The time required for various mating and testing operations also imposes constraints on the economic feasibility of programs. If the benefits of testing are sufficiently high and their costs sufficiently low, then two-stage selection procedures are feasible. These include collecting open-pollinated seed from intercrosses of prescreened parents, or clonal propagules of prescreened parents, or some mixture of these materials. A culling of the prescreened families or individuals follows testing, and the next generation of breeding is started from crossings among the selected seedlings or clones, usually in a seed orchard. The seed produced for commercial forestry use may be taken from a subset of the crosses, from the controlled crosses or open-pollination of all selected parents, or even from these in addition to other propagules, depending on their relative demand and supply. The costs of testing can be very high, and the added gain too low or uncertain, so the benefits must be well substantiated for any but the most important programs.

We are thus forced to distinguish among several kinds of populations that may be physically distinct. The major focus in this volume is the breeding population. The trees used for testing genotypic performance and the trees used in commercial forests may all differ from those used in the evolving germinal populations. We estimate the value of the breeding program on the basis of the breeding population, and can judge the efficiency of the various operations by its generation-to-generation improvement. In this chapter, we assume that the long-term breeding system is an iteration of the same breeding program. In later chapters, we depart from this form for the development of other breeding systems, but nevertheless can evaluate the progress for each breeding population by the generation-by-generation gains. We assume that a desired effective population size is predetermined and that a given number of parents is fixed for the breeding population.

3.3 Gain and Limits to Selection

The main purpose of selecting trees is to change the average performance of the population that is being subjected to selection in the desired direction. Many currently available publications specify the amount of gain that can be obtained from a single selection cycle of forest tree species. Most of these gains are predicted values determined on the basis of estimates of variance components and selection schemes used (Chapter 5). Because the generation turnover period of most commercial species is long and our breeding history is relatively short, breeders have rarely had sufficient time to determine realized gain figures. Nevertheless, the concept of expected gain (to be discussed in Chapter 5) has been used extensively to justify short-term breeding.

Similarly, long-term breeding is based on gain from selection as well as the need to maintain genetic variance. In fact, it usually takes more than one generation of breeding to significantly change populations. In subsequent chapters, we assume multiple-generation selective breeding for the purpose of discussing different gene actions, multiple traits, and multiple environments. Therefore, in the following section we review some well known results from recurrent selection (Section 3.3.1). As selection generations advance, however, the variance structure of the population may change, and the predictive power of parameters such as heritability may last only for a few generations, perhaps three or four (Hill, 1972). Therefore, the notion of using the gain, either predicted or realized, from multiple-generation selection as a parameter for precisely prescribing a recurrent selection system is unrealistic.

Similarly, the concept of limits to selection discussed in Section 3.3.2 is not an operational guide. It is unrealistic to establish a recurrent selection program with the idea that some day a larger population will reach a proportionally higher selection limit with respect to the traits of current interest. Limits to selection is useful as a conceptual restriction (Kang and Nienstaedt, 1987). For example, a small effective population size is one of many reasons why a population may fail to respond to selection. Understanding the concept of limits to selection may help the breeder to determine a "safe" population size, and, for a given population size, determine how the population structure might be manipulated.

3.3.1 Gain from Multiple-Generation Selection

For quantitative traits, the breeding record has indicated that repeated if sometimes uneven progress from recurrent selection is possible for many generations. Thus, if we connect the generation means and condense the time scale, we see that the cumulative progress indicates that the mean of breeding populations can well exceed the extrema of original populations in relatively few generations (Figs. 3.5 and 3.6).

Thus, with reasonably large population sizes for such quantitative traits as general yield, persistent selection over generations can result in cumulative gains. Even uncomplicated methods such as mass selection have obviously worked within traditional agricultural systems, and as the crops themselves changed in yield characteristics, concomitant changes in the cultural systems also occurred. Thus, as the crops and their environmental requirements become better known, performance under more predictable silviculture makes selection more efficient. It can therefore be expected that the heritability will rise somewhat because more accurate observation of yield-related traits will raise the effective genetic variance. It can also be expected that some environmental variables will decline as silvicultural regimes optimize environments or at least make them more predictable. Thus, field-stage screening can be expected to become more effective in the future. In the long run, there are biological and economic limits to how much environmental homogenization is efficient to force, and to how much variability is needed within tree varieties.

There are at least two ways for geneticists to breed for any given level of environmental variability. One is to breed for adaptability to an environmental range within a single variety or breeding population, and another is to breed for different environmental ranges in multiple populations. We postpone discussion of this topic to Chapter 6.

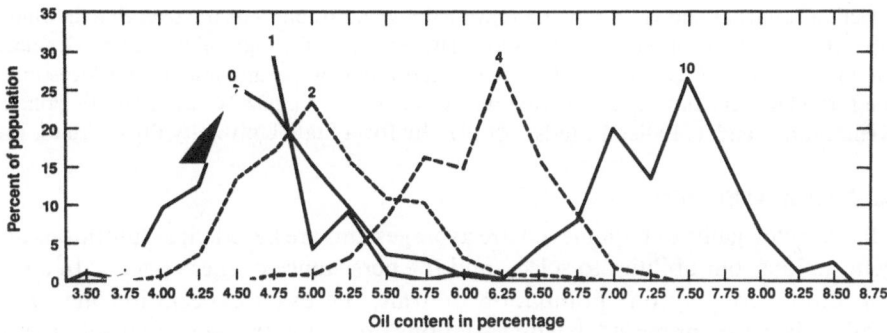

Fig. 3.5. Selection for high oil content in corn. Frequency distribution of original population is designated by 0 and the frequency distributions after various numbers of generations of selection by numbers. Mean oil content increased steadily with selection. (From Allard, Principles of plant breeding. Copyright 1960, John Wiley and Sons, Inc.)

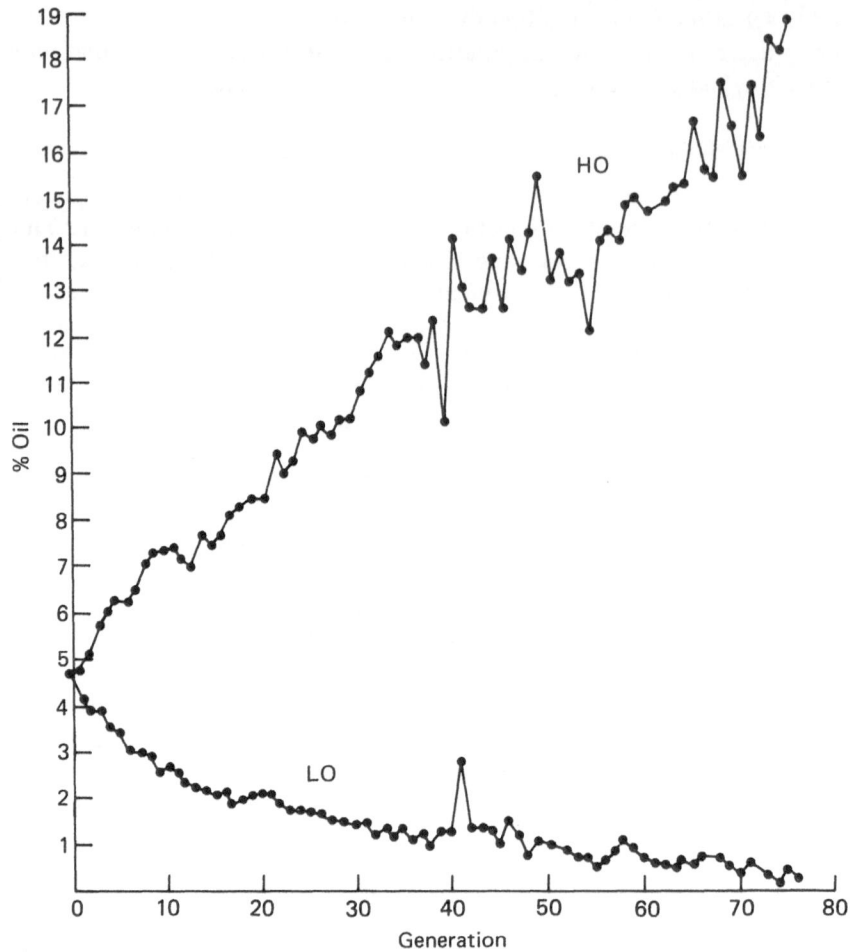

Fig. 3.6. Selection for oil content in corn after 76 generation of selection. Note that 10-year selection results as shown in Fig. 3.5 represent a small fraction of the overall gain obtained after 76 generations of selection. Symbols: HO, selection for high oil content; LO, selection for low oil content. (From Dudley, 1977. Reprinted by permission from Proceedings of the International Conference on Quantitative Genetics: August 16–21, 1976, E. Pollak, O. Kempthorne, and T. Bailey Jr. (eds). © 1977 by Iowa State University Press, Ames, Iowa.)

3.3.2 Limits to Selection

The expected gains to be achieved are averages and are based on assumptions about gene actions, our abilities to select, and the persistence of these effects. In any one generation, for any one population or trait, the exact expectation may not be realized because simple stochastic variations present in the particular set of parents or genes are effective in that generation. Furthermore, neither the selection differential itself nor the heritability may be exactly as expected. Thus, a variance in gain is to be expected, and small population sizes, or small numbers of genes affecting the traits, can make this variance large.

In addition to such vagaries inherent in any such predictions, other departures from expected responses may occur. Qualitative traits obviously do not display gradual changes if they are controlled by a few genes of large (qualitative) effect or even if controlled by many quantitative genes that must pass some threshold for their effect to penetrate to the phenotype. However, even for quantitative traits, the fixation of alleles ultimately limits the possible response to selection. As discussed in Chapter 2, the continued response to selection may be caused by many genes of small effect in populations of sufficient size that low-frequency alleles can continually be effective in generating a selection response, or epistatic effects can continually change the selection effect on allele frequencies such that genetic variations persist. Long-term responses to selection in plants (Dudley, 1977) and insects (Jones et al., 1968) indicate that even relatively modest breeding population sizes (e.g. 10–30) can continue to yield selection responses over many tens of generations (Fig. 3.6). These population sizes seem too small to allow us to consider that the persistence of very low frequency alleles can account for the continued response to selection, and hence it seems likely that mutations or epistatic effects are important in long-term selection response.

The altered physiology, anatomy, and morphology of organisms under long-term selection may force a type of epistasis in which allelic effects are changed not so much by differences in gene products, but more by the changed milieu. Thus, given an evolving environmental and genetic milieu, many alleles at many loci may ultimately be involved in the total cumulative gain achieved, while few may be substantially involved in any one generation. The problem for breeders in the long run is to ensure a reasonable chance for alleles of potential future value to persist in breeding populations, which requires maintaining sufficiently large population sizes in the collection of breeding populations. If the alleles of ultimate value have no positive or negative effects in the present or early generations of breeding, then no selection pressure for these alleles will exist, and their probability of ultimate loss is equal to their initial frequency. The population sizes required to keep the probability of immediate loss sufficiently low can be easily computed for any finite number of generations.

If the alleles of ultimate value have some persistently positive effects, then this positive selection pressure can increase the probability of ultimately saving and fixing these alleles. The population size then has an effect on increasing this probability, by a factor proportional to the $N_e s$, the product of the effective population size (N_e) and the selection differential (see Rawlings, 1970, for review).

If the alleles of ultimate value have negative initial effects, then initial selection will drive the frequency of these alleles down, and only if large populations are carried can low-frequency alleles exist through the exigencies of stochastic loss at lower frequencies. If this association of negative effects is purely the result of a negative linkage disequilibrium, then again large populations will tend to preserve alleles for a period of time that may allow for linkage equilibrium to be approached. The initial population sizes required to allow the ultimate probabilities of fixation to approximate a neutral effect model is an inverse function of the recombination rate. Thus, for loose linkage, (e.g., $r = 1/2$), a doubling of the population size over that which would be sufficient for early-generation allele survival would be sufficient for this case. Larger sizes are necessary for tighter linkages (Namkoong and Roberds, 1982).

Other problems seem to exist when new traits are added to the selection goals in long-term breeding lines (Dudley, 1977). It is unclear whether this is merely the effect of loss of alleles for these new selection criteria, or if the new selection pressures run counter to prior pressures and hence are likely to depend on alleles no longer present in the population. Indeed, reverse-direction selection experiments do indicate that allele loss from initial negative selection can be effective (Jones et al., 1968). However, it is also clear that large population sizes can maintain more useful variations for any changes in selection goals than small ones.

3.4 Developing Recurrent Selection Programs

In Section 3.2 we presented the basic selection model and some simple recurrent selection (SRS) models; and in Section 3.3 we discussed some concepts of interest in breeding that can be used in SRS and other models that will be developed in the subsequent chapters. We began Section 3.2 with an assumption that the species we are to be working with have already been chosen and that the appropriate source populations are known. Many programs have been successfully started from good first approximations. However, as we have learned more about breeding and the adaptability of populations, it has become apparent that we can adjust the site boundaries for planting genotypes, varieties, or species. Some special populations of one species might displace another species in some areas and in some generations, while some varieties may be so widely planted as to constitute all the planting of a species. Thus, initial estimates of species–site or provenance–site optimum pairings should be regarded only as first approximations. To handle such changes, there is a need to conceptually expand our simple breeding models (Section 3.2) even in cases in which the breeding capability is severely restricted by budgetary limitations. In this section we discuss some factors that vary the breeding circumstances. Each of the factors and means of coping with the situations are discussed in subsequent chapters.

For areas with a depauperate vegetation that may be caused by natural or man-made catastrophes, the initial phase of a breeding program is species and provenance testing. We discuss these programs in some detail in Chapter 8, but wish to emphasize here that our areas of concern in forest tree breeding must include the broader questions of how the breeding populations should fit into a more global view of forest management that includes uncertain and variable boundaries of adaptation. Thus, while there is a relational hierarchy from species to provenance to breeding population, it is not necessary to first demarcate species areas, then provenance areas, etc. However, as a first approximation, this ordering of concepts is a reasonable progression, even if it is often necessary to carry it out with meager data.

However, because programs are often started with little knowledge of the genetic structure of the species, there may be ways to select provenances or individuals and ways to structure breeding and production populations to avoid problems in changing an initial structure. For example, if individuals of a species have limited ranges of adatability, at least for some traits and for some environmental variables, then it would be inefficient or ineffective to breed for broad adaptability. The question of concern is which traits have limits and to which environmental variables. Some growth response traits for example may be highly responsive to soil, atmospheric, or temperature extrema, and hence some zonation is required for a set of different populations to provide an optimal, total forest area productivity. The trait may be survival against a pathogen environment, and again may require the use of several different resistance genotypes in populations.

Thus, if natural subdivisions of populations exist, as discussed in Chapter 1, or if genotypic mixtures exist in natural populations, such structures may be important for the survival of not only those "natural" populations but of planted forests as well. It is not necessary to mimic natural population structures, because we might have substantially different environmental variations and controls in planted forests than are considered in the evolution of the natural forest arrays. It is possible that we can homogenize an otherwise heterogeneous population, or make diverse an otherwise homogeneous population, but it seems obvious that if useful variations in trait responses or trait combinations exist we can make some use of preexisting population subdivisions. Thus, a deeper understanding of species genetic architecture can be useful in breeding, not only to retrospectively understand its sources of variation and improve sampling of its variations, but also to build on its structures for prospectively developing multiple varieties. This is discussed further in Chapter 6.

Similarly, if the heritabilities of important traits are low, careful testing programs can increase precision of estimated potential performance, and could increase our understanding of the environmental response functions that cause variations in performance. Thus, testing genotypes over controlled environmental arrays improves our knowledge of both the genetic control of response functions and the effects of environmental variations on performance. By such testing, the joint responses of multiple traits can also be studied and trait correlations analyzed for their genetic and environmental components.

All these advantages of increased knowledge for guiding breeding programs are of course foregone if no testing is performed. Thus, any self-correcting midcourse adjustments to breeding programs are also foregone in any breeding program that is reduced to the simplest elements of SRS systems. Then, even if future breeders may wish to forego these options of starting breeding populations with different provenance sources, of developing higher heritabilities, of adjusting for or using genotype–environment interactions, of breeding for different trait combinations, or of using different breeding methods for different gene effects, we would never know what opportunities are foregone without testing.

In the development of the initial tree breeding programs, breeders were faced with a dilemma: to either commit to a SRS program without testing (Fig. 3.3) or to develop SRS programs with at least some testing and estimation elements (Fig. 3.4). The former, SRS without testing, promised rapid initial progress and at moderate heritabilities could yield gains equivalent to programs instituted without progeny testing (Namkoong, 1970). The majority of early programs were started with at least some testing and estimation and have provided much of the information on which future breeding choices can be made. In some cases, larger and more diverse populations may be deemed best to develop, while in others, a reduction to very simple SRS programs may be preferred. Thus, an ongoing problem for breeders in all stages of program development is to simplify and reduce breeding population complexity as unnecessary for progress at least in the short run, or to continue research and development of potentially useful alternative populations. We discuss the design of the latter programs in Chapters 6 and 7, and indicate that many optional breeding programs can exist with different environments, traits, and gene actions. In those chapters, we discuss how to develop tests and populations to use such variations, but it is clear that the simplifying options always exist.

We may thus choose to never explore options for some particular species because our initial estimates indicate their potential value or amenability to development is low, or may even wish to homogenize other species into single breeding populations with homogeneous selection objectives at any time. On the other hand, we may choose to expand options in at least three dimensions. First, if there is genetic variation in environmental response functions, then it is possible that the maximum total forest yields would be best generated by breeding different populations for different zones. Second, if several traits jointly determine tree breeding values and are not inherited independently, then it is possible that breeding populations for different trait objectives is economically efficient. It is also possible, if trait values change in different environments, that we could breed most efficiently for particular sets of traits in particular environments. Third, if different kinds of gene action can be used for breeding, then some mixture of breeding methods, such as modified reciprocal recurrent selection (Mod. RRS) could be used. It is also possible that different breeding methods for different traits may be required to combine in a program of tandem improvement, and that different methods be used for different breeding zones.

The possibilities are numerous for organizing a breeding system, and are not limited to merely choosing high or low levels of intensity intended to homogenize a single breeding population with only one objective set of environments and traits and with only a single breeding procedure. The breeder will have to limit the species and populations for which complicated programs can be managed. Although not many programs may ultimately be treated so elaborately, the options for such developments remain unknown for many potentially useful species. Without some degree of testing, such possibilities may never even be imagined. Therefore, for forest trees, the extent to which variations in the breeding system are useful remains largely unknown. For most species, there will exist some limits of adaptability that are not well known. Thus, initial estimates of zonation or lack thereof should be considered as just that, initial estimates, and before finally assigning species to one or another level of breeding program intensity, the options should be explored.

The development of a species for wide adaptabilities can proceed in several ways, according to the extent to which ecological or economic demands vary and relative to the capacity for individual genotypes or populations to meet that variation. Thus, if economic demands are reasonably uniform and the environment is or can be culturally treated to be uniform with respect to individual adaptability, then single varieties can easily be bred for broad adaptability. Advanced breeding populations can then be developed for any temporal variations in ecology, pests, or markets, but with broad adaptability. Thus, even if some variation might exist in environmental response, such crops as tobacco have been tested and selected for uniformity over very wide geographic zones and for a uniform, high-input, cultural regime. Some food crops such as maize may also be developed by using very few lines in breed development for widely adapted varieties, while other crops such as rice may diversify local populations for adaptability to local conditions (Chang, 1985).

For most species, the question is, for how wide a range of adaptation can any single population be bred before substantial loss of local adaptability is felt and susceptibilities to economic or ecological shifts become dangerous. It is obviously simpler to create and test single populations for wide adaptability, but most such populations will eventually meet limits to their utility. For forest trees, the per-unit-product value is often low and this forces operations to be rather large scaled but usually not global in their extent. Thus, suspicions of limited adaptability have led to the drawing of seed transfer rules (Rehfeldt, 1984) and to breeding population zonations. We develop concepts of optimum zonations in Chapter 6, and at this time merely suggest that a conservative strategy would be to create zones as small as economic feasibility permits, and to test for breadth of adaptability. By such means, the options for developing locally adapted varieties are conserved but can be overridden if this is later deemed desirable. For species of low present but high potential future value, this may imply that low cost breeding options should be developed for limited traits and zones where breeding can have greatest impact. An alternative would be to develop a broadly adapted population for a low level but more general improvement effect. For long-term development, we prefer the first approach.

3.5 Vegetative Propagation

It is possible to vegetatively propagate individuals of several tree species not only for use in clonal seed orchards but directly as a means for generating plantable propagules. When this is an economically feasible alternative propagation system, several advantages accrue. Testing of exactly repeatable genotypes allows the breeder to observe the genotype that will be used in forest plantings in various conditions, and this can substantially enhance the useful heritability of the selection procedure. If testing can be performed with clonal propagules, then the error variance in estimating performance value can be minimized. This testing procedure is useful even if the forest planting system is to use seedlings generated from clonally tested parents.

If clonal stecklings (rooted stem cuttings) are to be planted, however, then the covariance between tester and testee is not diminished by the existence of any special clonal effects, and the full measure of all the genetic variance among stecklings is realized in the selection of clones. In clonal propagation, all the inherited variations contribute to the gain, whereas in sexual propagation, the gain is achieved on the basis of the interfamily portion of genetic variance. Thus, the full benefit of broad-sense heritability is realized rather than only some portion of the narrow-sense heritability. Furthermore, the tree breeder has full control over the amount of genetic variance to be allowed into a forest planting. By selecting one or a very few ortets, very little genetic variation would exist in the planting; by selecting for very diverse behavior, different patterns of genetically based variations may be included. Thus, the availability of such technologies as tissue culture, cuttings, and grafting, etc., bring opportunities for more efficient selection and control of genetic variation within plantings.

However, these advantages are often not directly realizable in a multiple-generation breeding program because the cumulative gains depend on allelic recombination between generations. Vegetative propagation merely multiplies selected genotypes, which must already be present in a population, and obviously cannot create new variations by allelic recombination. Cumulative breeding gains such as shown in Fig. 3.5 and 3.6, depend on new variations for selection whether dependent on simple additive gene effects or on more complex epistatic interactions, and sexual reproduction must be the basis for creating new variations. Thus, if the breeding program is directed to continuing improvement, then intermating among selected parents is required, and the cumulative gain will largely be based on the genetic variances described earlier in this chapter. If a form of Simple Recurrent Selection (SRS) is used, then the cumulative gain is based on the additive genetic variance and not on the other genetic variances, which might be the basis for a larger broad-sense than narrow-sense heritability. Thus, vegetative propagation systems are useful only within a short-term breeding program to provide an "extra" gain in each generation over that provided by the long-term breeding population. Over many generations, the rate of gain will be the same as that of the main breeding population, while the level of performance may be somewhat higher by a constant extra margin (Fig. 3.7).

In this sense vegetative propagation is not a breeding system, but is the propagation system used to produce forest plantings from the breeding program. Obviously, there are other ways to manipulate the breeding population, and by restricting the parentage of planted seedlings, the breeder can also obtain an extra gain, or can control the amount of genetic variation in the field. It is clear in these cases that the breeding and the field-planted populations are distinct, and that with either clonal or seedling plantings, the breeder may restrict or even expand the parentage beyond the limits of the breeding population. When constraining the number of clonal or seedling parents, there is a reduction in genetic variance in the field that may be an advantage if crop uniformity is desired or a disadvantage if diversity is required. A different danger exists when the loss of genetic variability is carried into the breeding population, because variability in this population is the foundation for future progress.

Reliance on vegetative reproductive systems often experiences a further drawback in that the method itself may impose curtailment of the selected population to those genotypes capable of regenerating in those ways. Thus, an extra "trait" is inadvertently imposed in the selection program, namely that of capacity for vegetative propagation. To the extent that it is limited to some genotypes, then selection

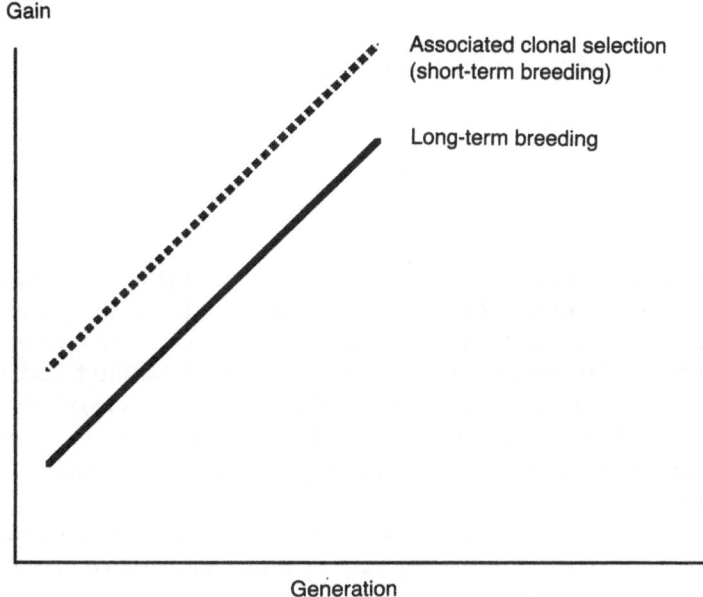

Fig. 3.7. Gain from long-term selection versus gain from clonal selection.

for other traits is diluted, or the population size is reduced. This is especially troublesome when different-aged trees are used and the decline in propagation success rates exists as a function of age. This is usually a more severe constraint than the ordinary sexual propagation biases that may exist in breeding populations, and extra efforts are required to avoid these detrimental effects on the breeding effort.

A problem in using vegetative propagules also exists in testing. If using stecklings to test but seedlings to plant (or vice versa), then the covariance of performance between test and planted materials is reduced by any independence between the physiological systems determining performance. Thus, if rooting capacity is independent of seedling growth, or if root morphology differs, then the performance of one propagule may have zero or even a negative correlation with the other kind of propagule. This is essentially similar to observing genes operating in different growth milieus, and can, but not necessarily will, cause problems. Such difficulties seem present in *Picea abies*, in which age seems to affect clonal performance and variations. Nevertheless, the techniques are sufficiently useful that they are regularly used in forest plantings (Rauter, 1982).

In any of these cases, the genotypic composition of the planted forest has come under close scrutiny as the possibilities of 'monoculture' have become technically feasible. Although this is only a concern for the temporary condition of one generation's forests, there is concern for the vulnerability of such genetically uniform stands to uniform disasters. This subject has been extensively discussed by Heybroek et al. (1981) and in a series of papers by Hühn (1985, 1986a, 1986b).

CHAPTER 4

General Recurrent Selection Systems

In Chapter 3, we defined a complete cycle of selection (Fig. 3.2) and developed simple recurrent selection (SRS) starting with a basic breeding system of simple mass selection in which seed is collected from the individuals judged best by their field performance. In this chapter, we consider more elaborate breeding systems with primary focus on the mating aspect of the complete cycle of selection (Fig. 3.2). In practice, all the components of the cycle including mating and selection must be considered simultaneously, but for present purposes we defer detailed discussion of selection until Chapter 5.

We begin this chapter by focusing on a single trait in a single environment. The trait may itself be a complex of many gene actions on several component physiological systems, but is one that presumably can be measured and conceived as a unit. The environment also may actually be heterogeneous, but because of our inability to understand the variables or to economically control the variation within a managerial unit, it must be treated as a unit with some inherent variation. Even for such a simple system, it can generally be expected that different kinds of gene action among many loci are responsible for trait expression.

In Section 4.1, we assume that gene actions are predominantly and persistently of the additive type, with average marginal effects that do not change rapidly as breeding populations advance. For this, simple recurrent selection is as effective as any breeding system. With this assumption we discuss different mating designs with respect to inbreeding, effective population size, and gain from multiple generation selection within the single breeding population.

In Section 4.2 we consider situations in which nonadditive gene actions, especially dominance, are also important. If intralocus interactions (dominance) are important or interlocus effects (epistasis) are large, then an individual's own performance may not reflect its potential value as a parent, and selections on that basis can prove ineffective. In the case of heterosis or overdominance, effective breeding requires that at those loci, the parents have different alleles, and in the case of epistasis, the multiple-locus combination of alleles, may change the breeding value from one generation to the next. When gene actions are mixed, then either a combination of breeding systems or a single compromise breeding system must be used. To effectively breed for nonadditive gene actions, more than one breeding population is needed, and matings in this sense represent interpopulation crossings such as used in hybrid breeding systems.

Although the concept of using more than one population designed to deal with nonadditive gene actions is introduced in Section 4.2, we still assume breeding for a single trait within a single environment. In Section 4.3, breeding for more than one trait and/or environment is discussed. The concept of multiple-population breeding is introduced and different concepts involved in handling more than one breeding population are discussed.

4.1 Mating Designs in Single-Population Breeding

In the first few generations of tree breeding, tests indicate that most of the genetic variance in commercially important quantitative traits is largely additively inherited. Even in those cases in which heterosis was considered to be involved, as in some species hybrids (Conkle, 1970), the evidence is strong that gene actions were largely additive. Similarly, in interprovenance hybrids, traits were largely additive and any nonadditive effect of hybrid performance seemed to result from the multiplicative effect of component traits such as survival and growth, each of which was additively inherited. Therefore, SRS would be sufficient for breeding trees for these cases, and inbreeding should not be a significant factor.

Although strong evidence of nonadditive gene actions is lacking for quantitative traits, heavy depression in such traits as seed germination, early growth and survival resulted from selfing and sib-mating (Franklin, 1968). While explainable by selection-mutation balance (Namkoong and Bishir, 1987), the overdominance hypothesis has not been entirely rejected (Bush et al., 1987). Observation of severe inbreeding depression in trees and the illusion of heterosis in some agricultural crops such as maize have served to counterbalance any wholesale adoption of SRS in tree breeding, and has kept alive interest in hybrid breeding (Box 4.1.1).

4.1.1 Intensive Inbreeding with Within-Line Selection

To develop an effective hybrid breeding program it is desirable to have highly inbred lines. Development of such inbred lines will, however, be impaired if viability or fecundity traits show severe inbreeding depression. Even if the breeder wished to develop SRS instead of a hybrid breeding program, the ultimate result of breeding within closed populations is a tendency toward homozygosis. Therefore, inbreeding depression on traits such as seed germination, early growth, and survival will eventually be expressed in the population. If the deleterious effects of inbreeding are the result of rare alleles at many loci (Namkoong and Bishir, 1987) that cause severe inbreeding depression, then purging these alleles from a population would prevent such severe inbreeding depression (Kang, 1982). The purging of deleterious alleles, however, is not a sufficient condition for developing high-yielding inbred lines (Box 4.1.2).

Box 4.1 Comments on Inbreeding and Heterosis

Box 4.1.1

Under natural mating systems with some degree of inbreeding, it is possible to observe an increase in yield or fitness under enforced outcrossing. However, depending on gene action and the degree of inbreeding, it is theoretically possible to observe an inbreeding or an outbreeding depression (Strauss and

---

(Box 4.1. *Continued.*)

Libby, 1987). Nevertheless, when a population base has been severely narrowed and inbreeding relative to some broader base population has occurred, it has been commonly observed that a boost in vigor is obtained on outcrossing. For example, the restoration of vigor to inbred lines of maize is well known, and it is suspected that *Pinus radiata* in New Zealand (Burdon, 1988) has increased vigor when crosses of its introduced and possibly narrow-base population are made to newly sampled sources. It has been suggested that a general level of heterozygosity is required for vigorous growth of trees (Ledig et al., 1983; Mitton, 1978). However, it is not clear that heterosis at the single-locus level, or cumulatively, is effective at many loci, or that breeding for heterozygosity in hybrid breeding systems would be effective in capturing any such effects (Bush et al., 1987). Inbreeding depression at the germination stage has sometimes been cited as evidence for heterosis, but can be more easily explained as the effects of a mutation–selection balance (Namkoong and Bishir, 1987). In a careful study of apparent heterotic effects in maize, the early estimates of overdominance gene actions turned out to be multilocus, epistatic effects with partial dominance at several loci (Moll and Hanson, 1984).

Nevertheless, there may be several loci that are important to maintain in heterozygous condition in any breeding or production population, and hence outcrossing among unrelated populations may be required for some commercial production. This might be achieved by maintaining sufficiently large population sizes in the breeding population, or by forced intercrossing among unrelated populations either deliberately developed for their hybrid effects or simply randomly chosen to ensure nonidentity of alleles at those heterotic loci.

Box 4.1.2

While purging deleterious alleles from the breeding population by an inbreeding program might prevent eventual inbreeding depression and/or help develop high-yielding inbred lines, matings among relatives, such as selfing or sib-mating, do not necessarily generate high-quality inbred lines. The ultimate worth of the final inbred lines depends greatly on the initial distribution of useful alleles in the progenitors. Thus, if useful variation exists at many loci, and any one individual progenitor has only a minority of the useful alleles, then even purifying the additive effects at the loci for which they have the beneficial alleles would not allow any useful recombinations to occur for eventual selection. Thus, inbreeding systems are dependent on the initial distribution of alleles, and if useful alleles are at all rare, then the existence of very good progenitors is improbable.

Box 4.1.3

Asexual means of haploid doubling would be quicker and would achieve homozygosity even if selection effects favoring heterozygotes at some loci are strong. However, these techniques have not thus far been successful with any forest tree species, and even in agronomic species, for which the technique seems to work, few useful results have thus far been achieved (Mayo, 1980).

Regardless of the objective, intensive mating of relatives, especially selfing in monoecious species combined with a within-line selection, represents a form of a complete cycle of selection (Fig. 3.2) designed to achieve a rapid approach to homozygosis while obtaining gain with respect to the trait of breeding interest.

In part because of the long generation turnover periods and in part because of the predicted difficulties with overcoming the initial severe depression with some traits related to viability of individuals (Orr-Ewing, 1965), this intensive inbreeding has not been considered as a useful breeding approach in tree breeding. A similar fear of inbreeding depression deterred animal breeders from aggressively developing highly inbred lines of livestock (Fisher, 1949). While such fear is justified with respect to short-term breeding, the same is not necessarily true in long-term breeding. If the deleterious effects of inbreeding depression are indeed caused by rare alleles at many loci, purging such alleles could become more costly in advanced generations than if they were purged early from the breeding population.

As is discussed in Section 4.3, selfing systems combined with within-line (or family) selection represent for monoecious species, a form of replicate population system with a population size of one. Assuming that the initial population is sufficiently large and that the offspring family size is large such that most of the lines can be prevented from becoming extinct, then the rate of initial progress of the selfing system could be higher than systems with larger population sizes (Baker and Curnow, 1959).

If the development of high-yielding inbred lines is the only objective for using the selfing system, then the long generation turnover periods and initial inbreeding depression of most commercial species will prevent tree breeders from considering selfing as a viable option. The combination of purging and the early response to selection, however, offer tree breeders a different possibility for using a selfing system, and the previously perceived constraints may no longer apply. The use of such a selfing system has been suggested as an addition to the main breeding system for breeding *Salix* in Sweden (Eriksson et al., 1984).

4.1.2 Regular Systems of Mating with Within-Family Selection

Progeny testing has persistently been a part of most tree breeding programs. As such, however, this testing has served primarily to make selection among the original entries more efficient. An important use of mating designs has also been to create a progeny population with particular family structures such that variance components can be estimated. For programs which then constructed the next breeding generation from intercrosses of the selected parents, the progeny are not used and the testing had no effect on the breeding system (Fig.3.4).

In a recurrent selection system, an important additional role of the mating design would be to create the progeny population that can serve in the breeding cycle as the basis for further selection. Some programs construct the next progeny generation from the unculled original progeny set. With incomplete intercrossing, some of the original parents would be unequally represented in the progeny generation. Still other programs construct the progeny generation without any control pollination of the selected parents, and some with only open pollination in seed orchards. For such mixed ancestries, the progeny population from which the next generation's parents are to be chosen does not contain all the gain possible to achieve and may contain a severely reduced effective population size (Box 4.2). Even in those control-pollinated populations, the demands of precise testing sometimes forced use of as few as four common parent testers, which severely reduced the possible effective population size of the progeny population.

Box 4.2 Effective Population Size

In Chapter 2, the concept of effective population size was introduced, and mathematical expressions for two different numbers, inbreeding effective number and variance effective number were given. Here we further discuss the effective population size in nonmathematical terms. This discussion is summarized from Kimura and Crow (1963) and Kang and Namkoong (1988).

The concept of effective population size was introduced by Wright (1931). The size itself represents the number in an idealized population in which each individual has an equal number of expected progeny. This idealized population size was used to understand the consequences of long-term selection of a finite population (Crow and Morton, 1955), especially in expressions such as Equation 2.42.

In real populations, however, individuals usually do not contribute equally to the next generation's 'gene pool,' assuming that the generations are discrete. Therefore, if we wish to learn the dynamics of the real population it is necessary to translate the census number of the real population to the corresponding number in the idealized population. As explained in Chapter 2, the key to the translation is to recognize that after a complete cycle of selection both the idealized population and the real population will generate a variance in gene frequency. Therefore, by finding the idealized population size that generates the gene frequency variance equal to that generated by the real population, the variance-effective population size can be determined (Equation 2.49).

It also turns out that, in any finite population, there is an inevitable increase in the inbreeding coefficient. Therefore, by equating the idealized population and the real population that generates the same increment in the inbreeding coefficient, we can determine an inbreeding effective population size (Equation 2.47).

Frequently the two effective numbers are the same and eventually converge, but they can be dramatically different. For example, Crow and Kimura (1970) considered

a 'population' consisting of one heterozygous individual which by self-fertilization produces a very large number of progeny. From the standpoint of inbreeding, the effective number is very small, for each pair of uniting gametes has a probability of 1/2 of being identical. On the other hand, the variance effective number is very large, for there will be very little random change in gene frequency; the progeny generation will have very nearly the same allele frequency, 1/2, as the parent.

This example implies that the number of crossings used in a mating design does not influence the inbreeding effective size. For example, a full diallel with multiple crosses will not necessarily generate smaller inbreeding coefficients

(Box 4.2. *Continued.*)

than that by one-pair mating using the same parents. The same is not true for the variance effective number. For example, Kang (1983) found that the variance of gene frequency change under family selection for single-pair mating (PM) is greater than that for partial diallel (PD). However, the allele-fixation probability of PM can be smaller than that of PD, even though the per-generation allele frequency change for PM is greater than that of PD. Both mating systems are balanced, but PD generates a larger number of families.

For breeding, it is also important to distinguish between the two different population sizes. First, the two numbers deal with different populations. The inbreeding-effective number deals with the parent (or grandparent) population while the variance-effective number is more directly related to the number in the progeny population. Second, the determination of inbreeding-effective number does not require the description of a complete cycle of selective breeding, because reference is made to the parent population. Variance-effective number requires the description of a complete cycle of selection with reference points at the time of reproduction (for example, see Robertson, 1961).

The implication for breeding of the distinction between the numbers is that the inbreeding-effective number is a most useful statistic for short-term breeding effects while the variance-effective number is useful for long-term breeding. In a strictly quantitative sense (i.e., ignoring the influence of inbreeding on survival of individuals, etc.), inbreeding does not influence the progress by means of selection, and inbreeding-effective number is not a factor in determining the selection limit. For example, Kang and Namkoong (1979) have shown that differences in the genotypic combination for a given gene frequency have little influence on the probability of allele fixation so long as the mating designs are balanced (i.e., variance-effective number was the same). On the other hand, for the same parental census number, some unbalanced factorial mating designs generated smaller fixation probabilities than single-pair matings (Kang and Namkoong, 1980). For a given census number of selected individuals and a given family size, the variance-effective number generated by factorial mating can be smaller than that by a single-pair mating because of the unequal genetic representation of parents in mating. We can then conclude that factorial mating is undesirable in multiple-generation breeding because it reduces the variance-effective population number, but not because of the related individuals it produces. In the short-term perspective, in contrast, if a seedling seed orchard is established from matings of selected parents the mating design used should not be factorial because the generated inbreeding-effective number is small.

It is possible to design a set of crosses for several objectives including both retrospective testing and prospective construction of breeding populations. For these purposes, mating designs obviously differ in the degree to which they permit the breeder to control the increase of inbreeding or coancestry for these dual purposes. At one extreme, the breeder may keep effective population sizes large by continually infusing individuals of unrelated ancestry, but if these individuals come from unimproved populations then there is a concomitant decrease in performance levels. The greater the infusion, the greater the decrease, and hence large population sizes and high gain rates can be achieved only if rapid selection gains can be obtained in order to enhance breeding populations. Alternative multiple-population breeding systems exist (see Chapter 5), but for single-population breeding, the essential closing of the population is needed to continually accumulate gains over generations.

At the other extreme, the breeder may close the population to a very few or even one parental genotype because that would represent the strongest selection possible. Family selection is effective at even moderately high heritabilities, and rapid gains can be achieved. Obviously, this would maximize current expected gain but would severely limit future gains (Sections 4.1.1 and 4.3). Many variations are possible between the two extremes, and for purposes of this discussion, we assume that the breeder makes an a priori decision on the desired effective population size and then chooses a mating and selection scheme. For a given number of parents (N), the effective population size (N_e) is maximized if each parent is represented in the next generation by an equal number ($2N$) of gametes. For most cases, we assume that N will be in the range of 20 to 50 monoecious individuals. This can be achieved by making N crosses, each parent being involved in two full-sib families. One such scheme is the circular half-sib mating design (Fig. 4.1a). As in the selfing system, gains from these kinds of systems primarily depend on within-family selection for any gain, because no parents are eliminated from representation in the next generation or are even permitted to be unequally represented.

At least in the initial generation of breeding, however, some loss or inequality of original parental contributions may be either forced or desired. The samples of the original population may be poor and require some culling, or the exigencies of breeding operations may require that some parental combinations be omitted. This often occurs, for example, when estimation and testing require that multiple crosses be made but full intercrossing of all entries is impractical. In these cases, if unequal gametic representation is desirable, then the number of individuals chosen must be larger than in the equal representation case in order to achieve the same N_e.

a)

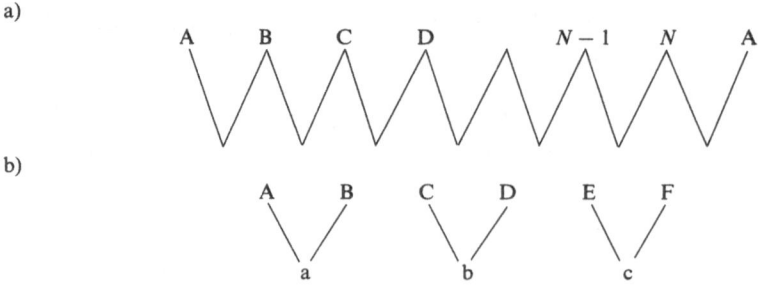

b)

Fig. 4.1. Regular mating systems with within-family selection. (a) Circular mating system. (b) Cousin mating system.

It is also possible to segregate subsets of parents, crossing only within a best set, within a second best set, etc., which would eventually lead to discarding the poorest sets. However, this would probably result in merely delaying a more stringent culling of the breeding population. On the other hand, mating schemes that pyramid the ancestries of entries (e.g., single-pair mating with monogamy and total avoidance of inbreeding) have also been proposed (Fig. 4.1b). Such schemes lead to a rapid buildup of coancestries by early avoidance of inbreeding, but eventually lead to a sudden jump in inbreeding, especially if family culling is practiced along the way. However, if more than one individual is chosen from each family in the first generation, and parallel mating schemes followed with these replicated family sets, then a system of intermating among the more distantly related sets can be instigated when the number of independent lines is reduced to one. These are so-called cousin systems of mating and also do not involve any early inbreeding. In comparison with the circular half-sib system, it does eventually catch up in the inbreeding coefficient.

All the above systems are forms of single-population simple recurrent selection, and if a major objective of breeding is to obtain both immediate and long-term gains, minimum single-population sizes must be maintained. The particular choice of breeding system may be more dependent on the costs and economies of controlling crosses and on the other benefits, such as testing, that might dictate the choice. Because most trees can produce large progeny families, an obvious expectation is that some regular system of crossing within closed populations will soon be established, with most of the selection pressure placed on within-family selection. In such designs as the circular half-sib system, for single-population breeding this could involve from 20 to a few hundred parents in each generation and several hundred progeny per family. Twenty parents might be considered a minimum to ensure some prospects of long-term advance in well known species; for most forest tree species that are only recently domesticated and for which many different kinds of traits may be of selective value, several hundred trees may be needed. As later discussed, this total may be subdivided into smaller units. As is the case with the selfing system, the possibilities of losing parental lines is great for small-sized subunits, because of the relatively common occurrence of inbreeding depression. Therefore for small subunits, for example, smaller than 4, it may become impossible to maintain the particular regular mating structure because of the loss of lines. Subunits sized between 8 and 12 may be feasible, at least in experiments on the effects of population size in breeding.

4.1.3 General-Purpose Mating Designs

Many different variations of diallel mating design and balanced factorial mating design fall in this group (Namkoong and Roberds, 1974). Instead of strict coancestry control, these mating designs simply require balance, and any design with equal representation of all parents can be established for other purposes such as estimating variances or for retrospective testing. Because these designs ignore relatedness, the selection and mating phases in Fig. 3.2 are independent of each other, and the only measure of inbreeding is the effective population size. Therefore, these mating designs can be applied to any type of selection.

This is commonly done in tobacco (Matzinger and Wernsman, 1968) and other crops in which complete intercrossing among selected parents is followed by individual selection without ancestral control. For general balanced mating designs, Kang and Namkoong (1979) found little loss in the fixation rate of desired genes in population sizes and mating designs commonly used in forest tree breeding. More relaxed control systems are also possible, such as using equal numbers of seed per female in the population developed for selection, but without pollination control. This might lead to highly unbalanced gametic representation (Müller-Stark, 1985; Friedman, 1981), and obviously if the control on equal seed contribution is relaxed, even greater imbalances can result. These can lead to rapid losses in N_e and would require a lower selection differential; or, for the same gain, a lower N_e would result. A loss of effective population size, and a higher probability of allele loss, can also be generated by unbalanced crossing, such as using a few parental trees as common testers (Kang and Namkoong, 1980). These mating designs also permit multiple crosses per individual if biologically feasible. This multiplicity, however, does not influence the effective population size (Box 4.2).

The particular choice of mating design, especially when multiple crosses per individual are easy to make, depends on the primary objective of making the crosses. In recurrent selection breeding systems with only additive gene effects and with every parent represented an equal number of times in the progeny generation, the pattern of crosses does not affect the expected gain. In such situations, the breeding value of each parent is included in the expected gain regardless of the particular mates it is assigned, and hence any mating is as good as any other. Then the choice of mating design depends on other considerations such as testing or estimation efficiency.

However, if there are useful nonadditive effects to be captured, then the specific crosses that provide heterotic or other nonadditive genetic gain opportunities would make it important to make multiple crosses. In the early stages of breeding programs when some parental entries are still being culled from large populations, it is also important to include multiple crossing opportunities for each parent. This would ensure that a good parent would not be paired only with a poor parent and risk being culled because of a poor mate. In both of these conditions, the breeder may wish to develop a hybrid breeding system that can cumulatively improve on the nonadditive gene effects, or may simply wish to use any available nonadditivity within a breeding cycle of a simple recurrent selection system. In either case, the more crosses that are made, the greater will be the opportunity of finding and using nonadditive effects. If some multiplicative gene actions exist such that there is a correlation between additive breeding value and specific combining ability, then a selective crossing of best × best parents provides a breeding value advance. Otherwise, the crossing pattern would be designed in ignorance of which particular crosses might be best and hence any random assignment of order in multiple crosses would be as good as any other, on the average.

In general, for breeding purposes in closed populations with no loss of parental representation in the progeny gametes, there is no need for balance as long as the individuals selected have equivalent representation of the parental gametes. In the circular half-sib system (Section 4.1.2), for example, the first generation could be generated by many possible connected designs, so long as the selected progeny can be drawn from a connected set of crosses (Fig. 4.2). Other crosses may be useful for

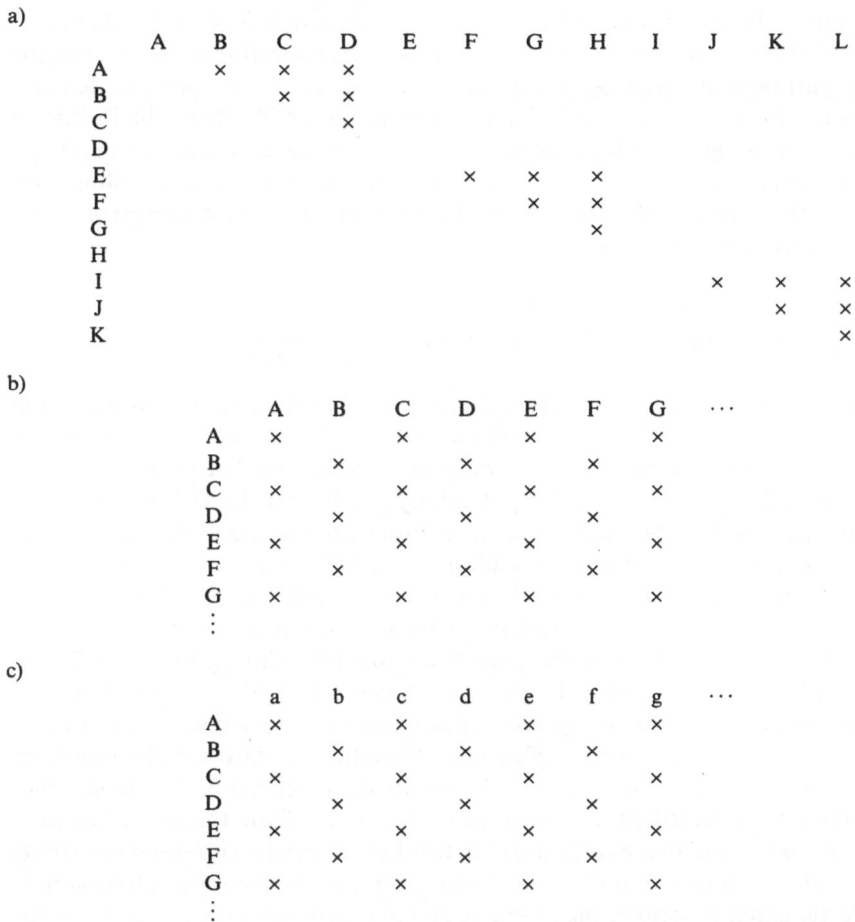

```
              A    B    C    D    E
        A          x
        B               x
        C                    x
        D                         x
```

Fig. 4.2. First-generation layout of circular half-sib system and cousin operation. Symbols: x represents circular half-system; *x* represents cousin system.

Fig. 4.3. Some variations of diallel and factorial mating designs. (**a**) Partial disconnected diallel mating design. (**b**) Partial diallel mating design. (**c**) Partial factorial mating design.

other purposes but are irrelevant to the breeding operation. In the cousin system, only every other cross (underlined) need be represented. However, in subsequent generations, the necessary subset of crosses is more proscribed, because the ordering of the parents is partially fixed by the first generation.

Thus, while it is necessary to maintain balance in the gametic representation from generation to generation, this can be accomplished within a subset of the crosses that may be useful for other purposes. For none of these purposes is it necessary to generate a complete set of intercrosses among all parents. Even for those cases just discussed in which it is desirable to make multiple crosses, there is a decreasing increment in the expected benefits as the number of intercrosses becomes large. Thus, some incomplete and often only partially balanced designs are needed. Obviously, neither factorial nor hierarchical designs, with a few parents crossing with many, provides a good basis for developing breeding populations, but many alternative designs exist that could serve multiple purposes. Partial disconnected diallels, partial diallels, or partial factorials could all serve (Fig. 4.3). In Chapter 5, we discuss these designs in more detail, for estimation and testing purposes. For the breeding purposes of this chapter, it suffices to note that the gametic balance requirements for maintaining effective population sizes still allows the breeder a wide choice of designs useful in forestry. Nevertheless, a clear discrimination of objectives for mating designs is useful to maintain. In addition to breeding considerations, the purposes of testing and estimation impose more constraints on the design of multiple crosses.

4.2 Reciprocal Recurrent Selection for Hybrid Breeding

If the choice of mating design for breeding purposes is not a difficult problem, and the choice of parental population size is soluble, then to construct breeding populations for recurrent selection we need only to consider whether simple recurrent selection is sufficient to capture the potentially available gain. Before discussing recurrent selection for hybrid performance, we briefly discuss recurrent selection for general combining ability that utilizes additive gene effects as developed for agronomic crops. Although these may not be especially useful in forestry at present, they might soon be considered as breeding populations develop more rapidly.

The term recurrent selection for general combining ability generally refers to crops in which a variety is already established as a standard and candidate lines are being developed to further improve those standard varieties. The testing of candidates is made with a mixture of gametes from the standard variety to observe the candidate breeding value against a broad varietal, genetic background. Once chosen, the set of selected parents are then intermated to form the next generation for recurrent selection. In the early stages of tree breeding, when no standard variety may exist, this method is not distinguishable from simple recurrent selection with several tester parents, because the objective and means of testing are to estimate the breeding value of parents against a general or wide sample of gametes from other potential parents.

In contrast, if there is a specific line or variety that is the known and fixed parent for which mating partners are to be selected, then it becomes the specific tester parent. On the basis of the test matings, the parents are chosen and then intermated to form the next generation for recurrent selection for specific combining ability. The breeding population in this case cannot be the same as the commercial variety, because the breeding population is the set of intermatings of selected parents whereas the commercial variety is a cross to the standard specified line or variety.

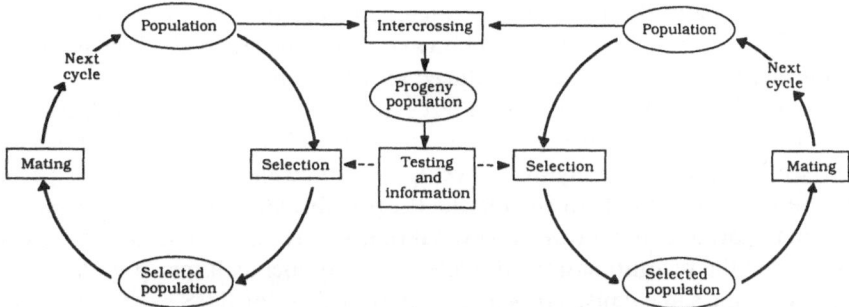

Fig. 4.4. Reciprocal recurrent selection cycle.

In the last case just mentioned, a general program for developing hybrid varieties has been assumed to have existed such that standard specific parents are already known. When this is not the case, as it will not be in most forest tree breeding situations, a program for developing both sides of hybrid parentage is needed. One method can be to essentially duplicate the simple recurrent selection breeding methods in two populations but to use the intercrosses between them to test for hybrid performance and for production of a commercial hybrid variety. The recurrent selection populations are maintained in separate, nonhybridized populations. The selected parents are crossed among themselves within each population to provide the next generation for recurrent selection, and are advanced with a presumably cumulative gain in their performance as parents of hybrids (Fig. 4.4).

The possibilities of long-term gain in this system obviously depend on the continued existence of useful genetic variation and hence require the same considerations of effective population size as do single-population, simple recurrent selection systems. Thus, the population sizes and mating patterns useful for developing the SRS breeding population are also useful for these systems within each of the parental populations. If alleles that exist at low frequencies are important to save and use in future generations, then each population may have to be as large as the one SRS population. However, if genes producing heterotic effects are important enough to warrant a hybrid breeding program, they may also tend to contain several alleles which exist at intermediate frequencies in the natural or base populations. If that is so, then each of the parental breeding populations need not be as large as those used for capturing alleles that are not held at intermediate frequencies. Then, the population size of each parental breeding population may be allowed to be smaller than that of a single SRS population that uses initially low frequency alleles.

One reason for developing the RRS method (Comstock et al., 1949) was to use both additive and dominance gene actions. Even if no dominance or heterotic effects existed, gain would be achieved if only because progeny testing would still be beneficial and selection for additivity would be as effective as in SRS. In fact, for most breeding populations, reciprocal testing is too large a program to handle, and some selection based on the individual's own performances or on pedigrees is used before entering the reciprocal crossing phase (e.g., Mod. RRS, Hallauer and Miranda, 1981). In forestry, such modifications would commonly be needed to reduce test sizes and duration. Thus, some gains are to be achieved by using the additive effects simply as a result of practical breeding necessities, while selection for heterotic effect may be made on a reduced set of parents.

It is possible to reduce population sizes and forego long-term progress, using perhaps only a few parents in each. The extreme would be to simply pick the parents of the best single-cross family and develop an inbred family line for the parental population. This would obviously restrict the possible allelic combinations to those present in the original parent(s), and even if repeated for many parental pairs would not allow for the recombination of alleles at many loci as larger population sizes would. Thus, the same constraints on population sizes for RRS exist as for SRS or line development. Single-line or inbred lines for single crosses are thus but an extreme form of reduced population-size breeding. Merely testing for parental "nicking" for specific combining ability is equivalent to merely picking the best of extant plants without recurrent improvement.

For a boost in gain within any one generation without any attempt at cumulative improvement in specific combining ability (SCA), the breeder can of course use just those specific pair matings for commercial seed production. However, unless there is some system for generating recombination in the parents, parental lines, or parental populations, there is no recurrent improvement. Thus, programs to use SCA such as two-clone orchards or strictly controlled crossing of specific parental pairs are not breeding programs, but are means of generating particularly valuable crosses for seed or clonal propagation. This is conceptually similar to picking good genotypes for clonal propagation, and may therefore be a means of propagating the best available material from any given breeding generation.

For traits with some additive and some dominance gene effects, the RRS and Mod. RRS will utilize both types of effects, but if the genes are completely dominant to overdominant in action, then selection on the basis of the individual's own performance is futile, and hybrid breeding is necessary from the beginning of the program. Thus, the Mod. RRS program, which uses additive effects in an initial stage, then uses both additive and dominance tandemly in the full RRS stage, might be less effective than a full RRS program. Even if gene action is all additive, in terms of the development of the breeding populations the RRS program will not lose potential gain, but it may be more costly. In particular, if the value of hybrids turns out to be the multiplicative or nonlinear effects of combining traits which themselves are additively inherited, then breeding for the traits in independent populations and combining them in hybrid populations will result in hybrid populations with the multiplied value of the two populations. This is, in effect, a form of independent culling level selection; as will be seen in Chapter 5, it is not as efficient as combined selection but will nevertheless yield gain.

The evidence from maize breeding, which showed initial hybrid vigor in yield, indicates that gain can be made with RRS, but that SRS can be as effective. In the experiments of Moll et al. (Moll et al., 1978; Moll and Hanson, 1984) two open-pollinated varieties, Jarvis and Indian Chief, were bred in two types of populations; the F_1 was used as a base population and a system of full-sib family selection initiated. In addition, a full-sib family selection system was started within Jarvis and Indian Chief and a RRS system was started simultaneously. Remnant seed for

Table 4.1. Results of using different recurrent selection systems in corn

	Grain yield		
	Generation		
Population	0	8	10
J	7.36	9.03	8.85
IC	8.54	8.49	8.92
RRS	8.93	10.77	10.84
J × IC	8.93	10.77	11.18
	Ear number		
	Generation		
Population	0	8	10
J	1.01	1.55	1.51
IC	1.38	1.50	1.73
RRS	1.29	1.75	1.83
J × IC	1.29	1.63	1.70

each generation was stored and periodically compared for progress. The progress as of the tenth generation is given in Table 4.1. As can be seen, both systems have made cumulative if uneven progress from their bases, with the RRS yields starting and ending at a higher level. As is evident from the yields of the pure populations defined from the hybrid of the pure Jarvis and Indian Chief populations but bred in a SRS system using full-sib family selection, comparable levels of gain can be achieved by selecting for additive effects.

However, as is evident from the yield of the RRS population over the initial hybrid and over the populations when bred within varietal types, gain was achieved by selecting for dominance effects also. Therefore, for these maize populations, similar gains could be achieved by using either additive or dominance genetic effects. For other populations or species, the balance could tilt either way, and it remains to be seen if hybrid breeding is especially useful even in maize. Because strong additive effects existed in this experiment, it would presumably have been possible to have captured most of the benefit of hybrid effects that derived from multilocus epistasis, simply by combining the varieties in the first generation and using the SRS system subsequently.

In forest trees, the question for breeders then is whether there is sufficient heterosis at the gene level to warrant a RRS program. Although it has been discussed as a viable option (Conkle, 1970) for interprovenance and for some interspecies hybrids of forest trees, there has been no strong development within any species. No recurrent selection program has ever been developed even for the hybrid poplar program, but it is not clear if this has been based on economic or technical decisions.

Some demonstrated hybrid superiorities may simply result from different loci, or traits, which on combination in an F_1 have nonadditive effects on the yield or other traits of interest. For example, the high-yielding hybrid between *Pinus rigida* and *P. taeda*, when planted in central Korea, owes much of its superiority to additivity in cold hardiness and in growth rate, such that it survives when *P. taeda* could not but grows at a faster rate than *P. rigida* could. An alternative strategy in this case would be to develop a SRS program using the F_1 population as a base breeding population for subsequent selections. It remains unclear that a hybrid breeding program is necessary for that case, and current breeding plans call for simultaneous breeding by SRS in the F_1 populations and for continuance of the parental population selections for a RRS program. Thus, even though substantial interest exists in the possible uses of hybrid breeding systems in forestry, the lack of established parental varieties for RS for SCA, and the necessity to develop two populations, have inhibited the development of hybrid varieties.

4.3 Multiple-Population Breeding

In discussing different recurrent systems we have encountered situations in which more than one population is deployed for breeding. For example, the individual line in a selfing system could be considered as a population of size one. The two sets in a cousin system and the two populations in RRS could be considered to be replicate populations. Although all such breeding systems involve multiple populations, the term *multiple-population breeding system* as used here represents a breeding system designed to handle multiple objective breeding, including the objective of adapting to multiple environments. To avoid the potential confusion between the use of *multiple populations* and the *multiple-population breeding systems*, we distinguish two more terms: *population copy* and *replicate population breeding system* (Box 4.3). Of the three concepts, multiple-population breeding systems provides the main focus of this section and subsequent chapters.

Box 4.3 Classification of Populations

Different populations used in breeding and related activities are classified in many publications (Burdon and Namkoong, 1983; Lindgren and Gregorius, 1976; Kang, 1982; Kang and Nienstaedt, 1987). For the present purpose we will restrict our discussion to three different concepts distinguished in the main text (Section 4.3): *population copy, replicate population breeding system*, and *multiple-population breeding system.*

Population copy represents the case in which the original ancestors of a breeding population are copied by clonal means, or progeny of the ancestors are equally divided into a number of genetically identical or very close groups. For example, Eriksson et al. (1984) proposed creating two inbreeding copies of an outcrossing breeding population by means of clonal propagation in addition to outcrossing a *Salix* breeding population in Sweden. The two sets of cousin system discussed in Section 4.1.2 represent copies. This definition indicates only the manner by which the replicate populations are created. In this case of *Salix* breeding, the copies are bred differently, while in the cousin system the two replicates are treated identically in the subsequent generations.

(Box 4.3. *Continued.*)

Replicate population breeding system and multiple-population breeding system primarily reflect the manner by which multiple populations are used in subsequent breeding. Although these terms do not indicate the means used to create the replicates, we can generally assume that the replicates represent nonidentical, or noncopy, samples of the ancestors. The samples may or may not be random. For example, jack pine breeding for the Lake States (Kang, 1980) represents randomly sampled replicates with each parent assigned to one breeding population, while a collection of SRS for white spruce breeding (Nienstaedt and Kang, 1983) would be nonrandom samples of the original ancestors. When the term multiple-population breeding system is used, the populations are not necessarily bred for the same trait or environmental adaptability objectives. As discussed in Chapter 6, an array of forestry objectives may require some differentiation in the genetic composition of the populations that breeders construct.

4.3.1 Multiple-Population Breeding Systems

If we consider breeding in multiple populations, the possibilities of breeding for heterotic effects and for trait combinations are somewhat easier to arrange than if the whole breeding effort was confined to a single population or population pair. In forestry operations, heterogeneous requirements or multiple objectives can be met in several ways. All requirements or traits may be combined into single trees, and all trees bred and managed the same way. Alternatively, a stand or managed block may consist of similar trees, but blocks may be designed to differ in their genotypic or silvicultural regimes. Another alternative is to mix different genotypes in single blocks so that different trees fulfill different goals but the block as a whole simultaneously satisfies several objectives. Obviously, the array of goals for each block may also vary, so that different mixtures would be present in the various management blocks.

For such a management system, it is conceivable that different breeding populations could provide genotypes for different objectives, which might include specific ecological adaptabilities, specific growth, yield, wood quality, and resistance types. Further, if gene actions are predominantly additive, then the breeding populations can generate hybrid populations that would be largely intermediate in the constellation of traits by which they differ. Such hybrid combinations of populations bred for different trait objective functions constitute a form of population level independent culling-level selection, which we discuss further in Chapter 5. As such, it is not as effective as index selection for all traits simultaneously for additively inherited traits. However, it is far easier to achieve multiple objectives by breeding multiple populations and using the hybrid for specific site-trait management blocks. If it turns out that the hybrid is better to breed as an independent population, then it would be easy to simply start a new SRS system on that basis.

Finally, the option always exists that if heterotic effects are found to be important for some traits, then maintaining the parental RRS populations would merely require the optimal pairing of parental populations. If, as conjectured, some form of Modified RRS is necessary to use, then the breeder may choose to select within each of the parental populations for the same or for different trait objectives. For example, if heterosis were strong for growth but not for resistance or wood qualities, the breeder may select on these latter traits independently in the parental populations or simultaneously in each.

4.3.2 Replicate Population Breeding Systems

An alternative form of multiple-population management is to subdivide single, large, breeding populations into smaller, independent units, with arbitrary initial assignment of individuals to groups. Breeding within each population is by some form of SRS, and for the same trait objectives (Baker and Curnow, 1959). By breeding in smaller units, some variation in effectiveness of selection would be expected to generate interpopulational variation in gain and variation in utilizing different alleles or loci to achieve those gains. Then, while the mean of the subpopulations might be somewhat lower than the expected progress of a single large population, the breeder could select the better populations to achieve a higher yield. Further, by intercrossing among these selected populations, the benefits of recombining different alleles of positive effect could be recovered for continued breeding advance. Intensive inbreeding with within-line selection as presented in Section 4.1.1 is an extreme example of the replicate population breeding system.

In spite of the uncertainty of the evidence with respect to ultimate advance, the replicate population method does make breeding somewhat simpler to carry out, and it is amenable to adjustment because smaller multiple populations can be easier to handle than larger ones. In fact, in early generations of tree breeding when a large degree of uncertainty about the ultimate structure of breeding populations can be tolerated, a system of subdivided populations might provide more flexibility for managing the breeding population. This is essentially what Lowe and van Buijtenen (1981) call *sublines*. In this proposal the populations are intended to maintain useful genetic variations within "lines," and no provisions for making use of interline genetic variations are presented. The primary utility of the sublines is that inbreeding can be avoided in seed production populations such as seed orchards, but not in breeding populations. It is similar in structure to the breeding programs for the British Columbia Forest Service Douglas-fir program, which contains six-tree units capable of being managed or maintained separately. The small number of subunits in that case was chosen to make the testing and variance estimation functions of the mating design simpler and clearer to institute, because partially balanced subsets can be fit into space and time constraints very easily (see Chapter 5). Such unbalanced designs are actually more efficient than balanced designs. They also lend themselves to breeding for multiple objectives, either for adaptability to different environments or ecological conditions, or for different traits for variable economic targets.

In fact, with some partial or complete control of pollination in seed orchards, the costs of single, replicate, or multiple population breeding are virtually identical. If seed production costs can be kept low, then any breeding system that can better fit genotypic distributions to more forest environments can extend the areas of breeding application. This has a critical impact on the economic profitability of tree breeding (Paques, 1984). The only major cost in applying even sophisticated breeding systems is intellectual. With cheap seed production, and widely distributed planting, selective breeding can substantially affect forest productivity even more than it has thus far. The only major questions are whether the variation should be managed with single or multiple populations, and whether selection should be for broad or more narrowly focused objectives. The efficient tree breeder can do either and can have a stronger impact on future forest management by focusing her efforts on developing populations to meet these evolving management objectives.

CHAPTER 5

Selection Techniques

In Chapter 3 we developed simple recurrent selection (SRS), which is as "good" as any other breeding systems provided that traits are influenced by additive gene actions, and that the traits of interest are not correlated. In Chapter 4 we relaxed the assumption of additive gene action and discussed means of dealing with non-additive gene actions. This is done primarily by changing the mating process of the complete cycle of selection (Fig. 3.2) and by introducing the multiple-population breeding concept. In this chapter, we further relax the assumption that the traits of breeding interest are either biologically or economically independent. To deal with multiple-trait breeding, breeders primarily change the selection aspect of the complete cycle of selection. Therefore we begin this chapter by discussing basic concepts of heritability, its estimation (Section 5.1), and expected gain from direct selection (Section 5.2).

In Section 5.3, methods of indirect selection, especially juvenile selection with respect to both short- and long-term breeding, are discussed. Sections 5.4 through 5.6 consider multiple trait-objective selection, which is the main theme of this chapter. Section 5.4 is a general description of three multiple-trait selection techniques: Tandem Selection, Independent Culling Level, and Index Selection. The utility of such selection schemes in forest tree breeding is discussed. Section 5.5 presents general strategies for handling multiple or varying breeding objectives. In Section 5.6 we discuss means of combining multiple gene actions and multiple-trait breeding. As the complexity of breeding increases, the problem of resource allocation arises, and means of organizing testing are also discussed in Section 5.6.

5.1 Heritability

In earlier chapters, we defined heritability as a measure of the importance of genetic effects in causing phenotypic variation relative to nongenetic or environmental causes. In the next chapter, we discuss the concepts and analyses of joint genetic and environmental effects on phenotypes. In this section, we discuss heritability concepts while assuming that there is a well defined set of environments, which themselves are of little interest but that collectively cause genotypes to behave consistently.

5.1.1 Heritability: Its Meaning

The assumption of independent environmental effects on heritability may often be a reasonably close approximation to reality as when many unknown microsite factors of the environment may vary randomly or in indiscernible patterns from tree to tree, with a net effect on phenotypes that is classifiable as random environmental noise which blurs the average genotypic expression in those environments. Because obviously there can be no phenotype without an environment with which the genotype finds expression, we do not say that the environmental variation blurs a genotypic expression, as if there were a disembodied genotype with a "true" expression. Rather, genotypes have phenotypic expressions in specific environmental sets.

Accordingly, even when there are well-known environmental factors, genetic effects can only be defined for those environmental sets. Thus, another source of inconsistent genotypic expression is when the environments may be known, but their frequencies may have an unknown distribution. For lack of good predictability, we may assume that some present distribution of environments is a good predictor of future environments, and therefore use an average genotypic performance to predict a general, average genotypic effect. In cases in which there may be a true random distribution of environmental effects, or a frequency or probability distribution of fixed environments, we therefore can often define an average genotypic performance. We defer discussion of deeper analyses of genetic and environmental effects to Chapter 6.

Variation in average genotypic performance, for some defined environmental set, is then an expression of genetic variation, and as a ratio of the total phenotypic variance, is an expression of heritability. Thus, for randomly sampled genotypes, the ratio of the variance among the phenotypic averages of genotypes to the total variance is heritability. In Fisher's fundamental theorem of natural selection, the genetic variance defined is the additive portion of the variance of average genotypic effects, and the rate of change caused by selection is proportional to this variance. For breeding programs, with an error variance around genotypic effects, the rate of change in a phenotypic scale is proportional to the ratio of genetic to phenotypic variance that is the heritability of the trait.

As described in Chapter 2 and detailed elsewhere (e.g., Namkoong, 1979), the expected phenotypic change is the product of heritability and the selection differential. Heritability therefore is useful as a concept for predicting gain from selection as well as an expression for describing populations with variations in genotypic effects. However, predictions of performance are often made on the basis of traits that may not be the exact same traits as those measured. For example, the traits as expressed in youth may not be the same traits expressed in maturity because the gene effects differ. Similarly, the biotic and abiotic environments of youth and maturity may well differ in their effects and hence the applicable heritabilities are dependent on the covariances of genetic and environmental effects.

5.1.2 Heritability: Estimation

Because heritability is composed of variances and covariances, its estimation usually involves estimating variance and covariance components. In the simplest cases, balanced genetic and field designs can yield clearly defined estimates of both genetic and environmental variance components. The standard designs, their analyses, and the interpretations of the variance components are reviewed elsewhere (Namkoong et al., 1966; Namkoong, 1979) and need no reiteration here. However, most tree experiments require extensive areas, which are expensive to establish and maintain and, by their blocking requirements, either incorporate large environmental hetero-geneities or prohibit testing on specific sites. The design and analysis of unbalanced designs for either variance component estimation (McCutchan et al., 1985) or Best Linear Unbiased Predictor (BLUP) and testing purposes (Friedman and Namkoong, 1986) for forestry are therefore significant advances in our capacity to improve selection.

By efficiently estimating variance components in such designs, the possibilities of much greater precision with smaller experiments is available. As beneficial as this might be in providing more precise estimates of variance components on given sites, of possibly greater significance is the expansion of sites available for inclusion in testing and estimation experiments. Because small replicate blocks can be used, large expanses of homogeneous planting areas are no longer required and two benefits are realized. First, smaller blocks on ridge tops or other special site types can be used, and second, even the larger contiguous planting areas can be broken into finer subdivisions of soil site types. The implication of the first effect is that site-response functions will be estimable over a more realistic set of environ-ments and, hence, not only will average environmental effects be better estimated, but genotypic variations in site responses will be estimable. If, by these new design and analysis methods, design constraints are substantially reduced, the forest geneticist can examine genetic effects over a more realistic span of environments, and we might be forced to reexamine what we mean by that "defined set of environments." We shall explore, in Chapter 6, the means by which the breeder may use genetic variations in response functions. For present purposes, however, it suffices to note that by including a wider sampling of environments, the environ-mental variance components will increase, the genetic variance might increase, but the net effect of heritability may either be an increase or a decrease even though the gene effects within any environment have not changed. By the use of BLUP (Friedman and Namkoong, 1986) with unbalanced designs, sites do not need to be as intensively managed and hence, may better reflect actual stand conditions. They are also much less expensive to operate and by reducing costs, breeding options that rely on testing such as hybrid systems and selection on low heritability traits may then become affordable.

Within a given span of environments, with a finer subdivision of blocks, the contribution of site variance effects to the error of estimating family means is decreased. Hence, for the same genotypic sources of variance causing family mean differences, the error variance is smaller and the heritability is increased. For the total span of environments we ought to consider in forestry, then, smaller unbalanced designs will yield heritabilities either larger or smaller than previous estimates depending on whether environmental sources of variance are added to or subtracted from the total variance. If multiple populations are bred for different environments, the environmental sources of variation would decrease within breed-ing zones and gain would increase.

It is also possible to construct very different kinds of heritabilities from the same experimental data but for application to different breeding situations. The genetic variance in the numerator is usually a genetic covariance between the traits and genotypes observed for selection and the traits and genotypes of the plants actually used (Namkoong, 1979), and may contain additive and nonadditive genetic variances. Thus, the numerator may be only a fraction of the true genetic variance of the trait for the environments tested and used. The phenotypic variance in the denominator may similarly be composed differently for specific breeding applications depending on the variances expected among the observed test entries, be they individual or families in single-observation or in replicated tests. We do not detail the combinations of numerator and denominator that are possible to construct for different breeding applications because these are available elsewhere (Namkoong, 1979), but merely wish to indicate that "heritability" is not a simple concept. Nevertheless, it does contain a consistent meaning for breeders as the heritable portion of the observed phenotypic variance that exists among the individual or family genotypic units of selection.

The possibilities of using clonal materials offers several more possibilities for estimating variance components as aids to selection, and as objects of selection. Obviously, for estimating the same kinds of variances as discussed, the use of replicates of the same genotype allows one to estimate another variance component, that is, between ramets of the same individual that would include only the environmental differences such as microsite effects and "m" effects from cloning (Park and Fowler, 1988). Obviously, by replication of an individual genotype the error of estimating an individual's performance can be greatly reduced, and hence the heritability for individual selection can be greatly increased.

5.2 Expected Gain from Direct Selection

Procedures designed to estimate the expected gain from several selection methods have been developed in Namkoong et al. (1966) and Namkoong (1979). A simple procedure is to merely select parents on the basis of their own performance, bring them into a clonal seed orchard or otherwise intermate, and to generate the next generation from a general intercrossing of all selected parents. The expected gain for this procedure is

$$E(\Delta G) = \frac{i\sigma_A^2}{\sigma_T^2} \tag{5.1}$$

This is similar to equation 2.35, but here σ_A^2 is the additive genetic variance, σ_T^2 is the total phenotypic variance, and i is the standardized selection differential.

A simple modification is to initially screen more parents than required for the previously determined effective population size, by the same procedures as above, and to progeny test before deciding on the final selected set of parents. Thus, the initial screening would include the same observed set of good parents as would have been chosen above, plus others that may, on first observation, not appear to be as good but which may rank highly on testing. The improvement is composed of a first stage with a smaller selection differential than above, and a second stage based on progeny testing, with a selection proportion sufficient to permit the same final effective population size to exist. The expected gain for this standard type of clonal seed orchard procedure is

$$E(\Delta G) = \frac{i_1\sigma_{A1}^2}{\sigma_T^2} + \frac{i_2\sigma_{A2}^2}{2\sigma_{PT}^2} \tag{5.2}$$

The product of the selection proportions for the two stages is expected to be often close to that of the selection proportion for single-stage selection. σ_{A2}^2 will be slightly smaller than σ_{A1}^2 (see Namkoong, 1979), and σ_{PT}^2 will often be slightly smaller than σ_T^2 because error reduction is the objective of progeny testing. Thus the gain from this two-stage selection will generally be greater than from one stage, but its direct costs and the burden of time will often be so substantial that one-stage selection will be more economically efficient. In future generations, the advantage of one-stage selection will be likely to be even greater (Namkoong, 1979).

An alternative procedure for the second stage of SRS is to use the progeny of the first-stage screening procedure to act as the parents for the next breeding generation or as commercial seed producers. The first-stage selection with phenotypic variance σ_T^2 is equivalent to the first-stage of Equation 5.2 if either selfed seed or intercrosses among selected trees are used, but is only half as much if open-pollinated seed is used. The second-stage gains with family variance σ_{PTF}^2 and individual variance σ_{PTI}^2 themselves are usually achieved by a combination of family and individual selection, and usually in tandem. The expected gain from such a procedure is

$$E(\Delta G) = \frac{i_1\sigma_{A1}^2}{\sigma_T^2} + \frac{i_2(1/4)\sigma_{A2}^2}{\sigma_{PTF}^2} + \frac{i_3(3/4)\sigma_{A2}^2}{\sigma_{PTI}^2} \tag{5.3}$$

In these cases, it is possible to derive an optimum allocation of selection intensities for any given set of heritability ratios, and hence to describe how large family sizes should be for within family selection intensity (i_3), the number of families to select (i_2), and the initial number of families to introduce (i_1). In addition, it is also possible to determine how the test designs can affect the phenotypic variances in the denominator of heritability, and thus an optimal design for maximizing gain in general can be found. Specific cases for *Pinus taeda* were analyzed, and the allocation of material was strongly affected by the relative costs of experimentation and the economic value of the gain (Pepper, 1983; Paques, 1984).

Whichever of these selection techniques is used, there is little doubt, with the kinds of genetic variances we apparently have for most traits of interest in most forest tree species, that progress has already been made and gains can accumulate in the future. Although the particular efficiencies of one method or another may differ depending on trait, species, and pollination and testing techniques, we shall probably be more knowledgeable about how, when, and where to measure breeding value, and more focused on which traits require more effort to measure. Thus, the basic breeding operations seem to be well established even if still being developed. We might therefore expect that with reasonable care in maintaining N_e, and better techniques for early estimates of performance and inducing early reproduction, that the amount of gain can continue to accumulate for many generations with a possible shortening of the time required to cycle generations. We might therefore envision the progress of subsequent generations of breeding along a time scale as a sequence of population distributions with improving means such as are shown in Figs. 3.5 and 3.6.

5.3 Indirect Selection

Direct measures of trait performance on the individuals that may be selected for breeding is the simplest concept of selection on gene effects. In contrast, observing a correlated trait on a relative that may be of a different age and grown in a particular environmental regime is obviously a far more indirect way to select. Nevertheless, it is often opportune to select on the basis of the performance of other individuals and of other traits. Furthermore, it will become apparent that we must generally consider that selection on any set of traits carries responses in other traits and, therefore, any selection requires a broader consideration of how to best handle multiple-trait responses.

The idea of using relatives as observable phenotypes for selection is, of course, neither new nor unique to forestry. The concept of pedigree selection as used in crop breeding involves selecting parental lines or remnant seed of plants chosen on the basis of their progenies' cross performance. While this is specifically applicable to selecting inbred lines for hybrid performance, it is in spirit not unlike other forms of progeny testing in which the progeny may be more easily measured, more easily grown in the environments or over the environmental range desired, or of a more appropriate age and condition to test. Some traits may only be expressed in one sex or at one particular age, and if the selected trees are of another sex or age, then the trait may be unobservable or so poorly measurable that selected and tested trees must be different.

Although rarely as difficult as testing for milk yield in selection of bulls, sexual differences in dioecious trees can cause problems. Age differences often substantially affect performance and hence older trees may not reflect juvenile performance, while juvenile performance may not foretell mature tree behavior. Similarly, testing for performance in special or stress environments may require using seedling or clonal replication, which may be relatives of the parents to be used in breeding, in several environments.

The most common use of aids to selection has been of progeny to help evaluate parents, but any of several kinds of relationship can be used. Older parents of trees that may be bred for traits observable in more mature trees can reverse the usual relationship, and if grandparents are also observable, they may further add to the performance data. Siblings are also often used in seedling seed orchards, and in some second-generation selections, full- and half-sib family means are used to aid selection.

5.3.1 Estimating Correlations

In all these cases, any trait expressed at one age, sex, or environment may not involve the same genes as in another condition. In any event, the genetic covariance is symmetric with respect to traits and relationships among individuals. Thus the breeder can select on one set of traits in one relative, and affect another set of traits in another relative, and can also reverse the procedure, but with unequal effect because selection response in the two cases reverses the roles of dependent and independent variables.

Another source of phenotypic correlation is environmentally based and independent of either of the genetic sources of correlation. One environmental source of variation, for example, soil fertility, may result in an increase or decrease in both height growth and resistance simultaneously and, without adjusting for it, the phenotypic correlation would be a biased estimate of the genetic correlation. Obviously, different environmental factors can result in either positive or negative correlations. In addition, there may also be a correlation of environmental effects, that is, a correlation in the occurrence of forces that independently affect traits, rather than an "environmental pleitropy." Phenotypic correlations, like genotypic correlations, can result from the frequency of occurrence of an environmental parameter that promotes the development of one trait and co-occurs with another environmental factor that promotes or inhibits the second trait. Either type of environmental correlation may be managed by some site interventions, but it behooves us to separate the environmental types of correlative effects among themselves and between them and genetic effects. Although a primary objective of many analyses has been to separate the genetic and environmental sources of variance, it may be almost as important to distinguish among types of sources.

Finally, correlations at different levels of genetic organization may also differ. We can imagine for example that, at the level of individuals within families, or families within varieties, a negative genetic correlation may exist between growth rate and certain wood density, phenol concentrations, or resistances. However, among different varieties or species, large plants may be associated with longevity, wood production, or different phenolic production such that positive correlations appear. It is of course not necessary for the correlation to switch and in fact it may often not do so, but it is conceptually feasible. Similarly, for any set of environmental factors that may be hierarchically or otherwise grouped, the correlations from microsite variations may be quite different from those from general or more global site variations.

We therefore envision a need to distinguish among various sources of correlation and to define the components of covariance that are included in these correlations. If we first assume a single level of genetic organization, and a single environmental effect, then an analysis of covariance can provide estimates of the strengths of the source of covariance. Using traits x and y, where each is a linear function of genetic and environmental effects,

$$x_{ij} = \mu_x + g_{xi} + e_{xj} + \varepsilon_{xij}$$

$$y_{ij} = \mu_y + g_{yi} + e_{yj} + \varepsilon_{yij}$$

The covariance of two observations is

$$E(x_{ij}y_{ij}) - E(x_{ij}) - E(y_{ij}) = E(g_{xi}g_{yi}) + E(e_{xj}e_{yj}) + E(\varepsilon_{xij}\varepsilon_{yij})$$

$$= \text{Cov}(g) + \text{Cov}(e) + \text{Cov}(\varepsilon)$$

assuming no covariance between genotypic effects and environmental effects or error deviations.

In an Analysis of Variance (ANOVA) and an Analysis of Covariance (ANCOVA), we may tabulate the expected mean squares and mean cross products as in Table 5.1. The ANOVA is done for each trait, and the ANCOVA for each pair of traits. In those cases in which there is no question of the independent measures of the two

Table 5.1. Expected mean squares of an ANOVA and ANCOVA[a]

Source	EMS	
	ANOVA	ANCOVA
Family	$\sigma_e^2 + E\sigma_f^2$	$\sigma_{e_1 e_2} + E\sigma_{f_1 f_2}$
Environment	$\sigma_e^2 + F\sigma_E^2$	$\sigma_{e_1 e_2} + F\sigma_{E_1 E_2}$
F × E (error)	σ_e^2	$\sigma_{e_1 e_2}$

[a] σ_e^2 is the error, σ_E^2 the environmental, and σ_f^2 the family variance component; $\sigma_{e_1 e_2}$ is the error, $\sigma_{E_1 E_2}$ the environmental, and $\sigma_{f_1 f_2}$ the family covariance component; and E is the number of environments, and F the number of families tested.

Table 5.2. Expected mean squares of an ANOVA and ANCOVA with plot error terms[a]

Source	EMS	
	ANOVA	ANCOVA
Family	$\sigma_w^2 + n\sigma_{fe}^2 + nE\sigma_f^2$	$\sigma_{w_1 w_2} + n\sigma_{fe_1 fe_2} + nE\sigma_{f_1 f_2}$
Environment	$\sigma_w^2 + n\sigma_{fe}^2 + nF\sigma_E^2$	$\sigma_{w_1 w_2} + n\sigma_{fe_1 fe_2} + nF\sigma_{E_1 E_2}$
F × E	$\sigma_w^2 + n\sigma_{fe}^2$	$\sigma_{w_1 w_2} + n\sigma_{fe_1 fe_2}$
W (error)	σ_w^2	$\sigma_{w_1 w_2}$

[a] σ_{fe}^2 and $\sigma_{fe_1 fe_2}$ are the family × environment variance and covariance components. σ_w^2 is the within plot error variance component. $\sigma_{w_1 w_2}$ is the within plot error covariance component. n is the number of trees per plot.

traits, the interpretation of the covariance is straightforward. However, in some cases, such as examining the same trees in a time sequence, there is a question of the independence of the error terms and hence of the covariances because they may confound subtle environmental covariances. For example, if trait A is early growth, and B is the same trait at just a little later period, then A may be either a component of B or the error and environmental effects in B may be entrained by the effects of A. If A is merely an additive component of B or is an ontogenetic precursor of B, then it may be legitimate to call the effects that A has on B's environment a part of the genetic covariance. However, if the breeder intends to use trait A to estimate trait B in independently growing individuals, then that correlated effect should not be called a part of the genetic correlation. In that case, to separate these kinds of subtle but biasing effects, independent trees must be used to estimate the covariances. For an analysis that extends to a within-plot error level see Table 5.2.

By simply using different seedlings or clones for each measure it is possible to estimate these covariances clearly. Then, the interpretations of the design covariance components in terms of environmental and genetic sources of covariance are exactly equivalent to the interpretation of variance components.

It is now a simple matter to construct various correlations. For example, the additive genetic correlation can be estimated if the families are half-sibs and we ignore epistatic effects. This is merely the additive genetic covariance for traits 1 and 2, divided by the geometric mean of the additive genetic variances for each trait:

$$r_{A_1, A_2} = \frac{\sigma_{A_1 A_2}}{\sqrt{\sigma_{A_1}^2} \sqrt{\sigma_{A_2}^2}} = \frac{4\sigma_{f_1 f_2}}{4\sqrt{\sigma_{f_1}^2} \sqrt{\sigma_{f_2}^2}}$$

A microsite error correlation for traits 1 and 2 can be estimated as

$$r_{e_1, e_2} = \frac{\sigma_{e_1 e_2}}{\sqrt{\sigma_{e_1}^2 \sigma_{e_2}^2}}$$

or by extracting the additive genetic contributions contained in the error components, and assuming no dominance genetic effects, a microsite environmental correlation can be estimated as

$$r_{\varepsilon_1, \varepsilon_2} = \frac{\sigma_{e_1 e_2} - 3\sigma_{f_1 f_2}}{\sqrt{\sigma_{e_1}^2 - 3\sigma_{f_1}^2} \sqrt{\sigma_{e_2}^2 - 3\sigma_{f_2}^2}}$$

A family correlation can be estimated for the given experiment as

$$r_{F_1, F_2} = \frac{\text{MCP}(F_1 F_2)}{\sqrt{\text{MSF}_1} \sqrt{\text{MSF}_2}}$$

$$= \frac{\sigma_{e_1 e_2} + E\sigma_{f_1 f_2}}{\sqrt{\sigma_{e_1}^2 + E\sigma_{f_1}^2} \sqrt{\sigma_{e_2}^2 + E\sigma_{f_2}^2}}$$

Another possibility is to define a correlation for individuals in different families grown in the same plot environment as

$$r_{\text{ind}}(1, 2) = \frac{\sigma_{e_1 e_2} + \sigma_{f_1 f_2}}{\sqrt{\sigma_{e_1}^2 + \sigma_{f_1}^2} \sqrt{\sigma_{e_2}^2 + \sigma_{f_2}^2}}$$

A total correlation can be composed simply by summing all components of variance or covariance that would apply to random individuals grown in random environments. Obviously, what is meant by a genetic correlation can vary substantially, and its estimator could include any of several types of components with various coefficient weightings.

Although rarely done, it is possible to consider further partitions of the genetic covariances into covariances of dominance and epistatic terms, the covariance of additive effects in one trait with dominance effects in another, etc. Perhaps of greater practical significance would be the partitioning of the environmental sources of covariance into covariances from any of several environmental sources as obtainable from factorial experimental designs. For example, the correlation between height growth and wood density in pines might be affected positively by general growth-promoting factors such as soil fertility, but negatively affected by spring rainfall levels. The effect on the genetic correlation may also vary if gene effects vary over these environments and hence a g × e effect in correlations may be generated. Such analyses among different environmental factors and several levels of genetic factors could be very enlightening of the physiology of correlated trait development of forest trees. Clearly, in the composition of a correlation that includes positive and negative covariance components, the size and magnitude of the correlation depends, either implicitly or explicitly, on the particular mix of factors included. Indeed, one possible source of variation in the estimated genetic correlation between generations may not be genetic at all, but rather a change in the environmental effects on the covariance.

In general, the designs that provide efficient estimators of variance components will also provide efficient estimates of their covariances as well. However, the correlations contain three times as many components as simple variance estimators do, and hence are estimated with relatively large error.

In general, the design problem is not only to estimate the several variance components, but also to estimate the covariance components as well. Thus, instead of only estimating a genetic and environmental variance, as in Table 5.1, the objective of estimation is a whole matrix of genetic variances and covariances and a comparable matrix of environmental variances and covariances. Thus, even with just two components to estimate, we are faced with estimating two matrices with $n(n + 1)/2$ elements for n traits.

5.3.2 Juvenile Selection for Short-Term Breeding

One of the more economically important uses of correlated trait selection in forestry is the estimate made of mature tree performance by the tree, its sibs, or its progeny when they are young. This can be reversed, as when older parents are selected to generate progeny that perform well at younger ages. However, the economic advantages of being able to observe traits in young seedlings, to reduce direct costs, and possibly shorten the generation interval are often great enough that juvenile selection becomes highly desirable. In this selection, it is generally assumed that the same genes and processes active in early and later ages are largely the same and hence juvenile expression is highly correlated with mature tree expression. This is rarely exactly true, because both the physiological system and gene expressions themselves change with the accumulation of size and years. Juvenile growth processes are not exactly the same as those in adult trees, and they operate in a different morphological/physiological state. Nevertheless, some growth processes may be identical, and some traits may foretell what later behavior will be like even if they are not identical.

To realize the benefits of early selection, early test data must be available when needed and hence the timing of the testing and selection phases must be coordinated. Thus, if progeny or sib testing is excessively delayed, selection may have to proceed before even the juvenile performances are observable, or if selection cannot be delayed, then progeny selection might be made on the basis of adult performance. In fact, in the initial generations of tree breeding, it can be expected that some inefficiencies will be experienced as more is learned about the genetics and physiology of trait development and as correlations become better estimated.

In the first generation, breeders may not be willing to trust the persistence of trait correlations and hence such correlation studies will be more effective for improving efficiencies in subsequent generations. When studies on juvenile–mature correlations are just begun in the first generation of breeding that information will generally only be available when the first generation is mature and too late to benefit that generation. A question is then raised about the stability of the correlations if they are generally to be estimated on materials one generation removed from use. That is, we estimate the correlation on trees in generation 1, but only have that estimate when generation 2 is just beginning, and may not be assured that the correlation for the adults of generation 2 will be the same. Therefore, less reliance can be placed on traits when the correlation is the result of ephemeral environmental effects, linkage disequilibria, or other causes that are not physiologically entrained.

If trait performances are consistent over years, or if juvenile performance is directly the economically important trait, then we have no problem in direct observation. However, if we wish to estimate performance at age x, then correlation of trees aged $x - 1$ will generally be better than that of age $x - 2$, etc. Age or physiological thresholds may cause sets of ages to be similarly good or poor, and periodic behavior may require long times to observe the trait well. Aside from these, however, the general expectation is a steady decline in the genetic covariance as departures from age x increase. For some period of time, near x, the same genes may be expected to be operating within the same environment, and hence the covariance will be nearly equal to the genetic variance, and the genetic correlation will be nearly at its maximum of 1. However, at earlier ages, a different set of genes may be operating; if these are the same genes, they would be operating in a different external and internal milieu. The genetic and physiological interactions may be so different and dependent on other gene effects, that an essential independence of gene effects "on the same trait" may exist.

Nevertheless, some traits will be found to be highly genetically correlated in different ages whether they have the same name or not, and if the earlier expressed trait has a high heritability, then juvenile selection may indeed be consistently good. It is these kinds of correlations that make juvenile selection most profitable, and which should be sought, not only by pairing traits of the same name, but by looking for any fortuitously high correlations. While most of those ontogenetically unrelated traits may be found to be merely in linkage disequilibria, some may give clues to common physiological effects of pleiotropic genes.

The question for the breeder is the time at which a given level of correlation is sufficient to justify selection. In most breeding programs, no selection is done before the parental trees reach mating age. This might be only a few years if parents in clonal seed orchards are to be selected on the basis of juvenile progeny, or may be 15 or more years if seedlings are to be grown to sexual maturity. Even then, if the correlations between 15-year and mature tree performances are poor, the breeder can choose to delay selection until the correlation is high enough. It is sometimes advantageous to select even before sexual maturity of the selected parental trees if some operational costs can be reduced by early culling, or if special measures can be used on selected seedlings for early sexual maturity. Although the costs for carrying tests and for delaying the breeding cycle vary considerably, and the economics of discounted value affect the benefits/cost evaluations, Nanson (1970) estimated that in general surprisingly low correlations can justify very early selection.

Thus, even correlations as low as 0.6 can be useful if the breeding cycle can be shortened, and if the reproductive technology is available to take advantage of early selection. For this reason, an overmeasurement of possibly correlated traits may be worthwhile simply to scan as many traits as possible in young trees for later performance. The estimation procedure may not only involve observing many youthful traits for correlation with performance of those trees in later years, but can also include correlation estimates of older parents on youthful progeny, because the correlations are symmetrical.

In the spirit of targeting traits for closer examination of their potential associations between juvenile and mature tree growth, Franklin (1979) hypothesized that the low correlations between seedling and mature tree height growth observed by Namkoong et al. (1972) and Namkoong and Conkle (1976) were the result of change in the competitive environment. He therefore proposed that crowding juvenile trees would mimic the competitive environment of older trees at wide spacing and hence boost the correlation. This proved questionable in the case of young Douglas-fir (Campbell et al., 1985). Other attempts to select particular features of juvenile growth to discern a mature tree growth capacity have focused on the shoot-to-root ratio, measuring the "determined" stem growth and growth rhythms instead of the primary growth of seedling (Ununger, 1987), and CO_2 photosynthetic efficiency (Ledig and Perry, 1967; Ledig and Fryer, 1972).

Such searches have developed a wealth of information about the phenomena that make up the growth process and may be integrated into a good estimator of growth potential. The unanswered question at this time is, if the active genes themselves are the same, is their effect the same, and can they operate in the environment of a forest of mature trees as they do in seedlings? If so, then to be better breeders we have only to be better physiologists. If not, then there is another problem in prediction that is not soluble merely by searching for well correlated traits.

There is a further problem in using correlations involved in the substantial problem of estimating correlation coefficients. Because correlations are estimated as a ratio of estimated variances and covariances, the error of estimate is relatively higher than the error of estimating a single component of variance, and to ensure that the family mean genetic correlation is in fact greater than some minimum requires experiments with many tens of families.

Because the genetic causes of the correlations may be changing over the generations of a breeding program, estimation experiments must be continually made, and breeding goals may also require adjusting if correlations switch from positive to negative. In maize breeding experiments, for example, the correlations between yield component traits are notoriously volatile even when measured with reasonable precision. Because genes do not act independently of the physiological or environmental system, it might be useful to ask if any such changes in the actual correlations carry information about the effects of genes on physiology or if gene effects are environmentally dependent in the way in which they operate. Thus, for example, a change in growth-related juvenile–mature correlation might indicate the existence of differential growth physiology or a change in which genes are simultaneously affecting the two traits. It would then be highly informative and of potential benefit to study the genetics of development, and to not merely rely on new estimation experiments.

5.3.3 Juvenile Selection for Long-Term Breeding

In short-term breeding, breeders work with the genetic correlation that exists in the population at the given time, and hope that it does not change drastically in one or two breeding cycles. Their main goal is to determine the time of selection that would maximize the annual gain from selection with respect to the mature trait.

In long-term breeding, the major question is how to handle potentially changing genetic correlations over many selection cycles. As is the case with inbreeding, breeders can take two opposite views on this subject. They could look for means to break or substantially change the correlation, or look for breeding techniques designed to maintain or increase genetic correlations. In either case, it is necessary to understand the dynamics of the correlation.

We have already mentioned that genetic correlations arise from either linkage disequilibrium or pleiotropy, or from both. Two simple models generating genetic correlations are shown in Box 5.1. In the past, pleiotropy received more attention than linkage disequilibrium as the cause of genetic correlation. Both Lerner (1958) and Falconer (1981) used pleiotropy models to intuitively demonstrate the potential behavior of genetic correlation under selection. Lande (1980, 1984) suggested that pleiotropy might be more important than linkage disequilibrium in maintaining genetic correlation. This view, however, is concerned with changes in correlation on an evolutionary scale, and linkage disequilibrium remains an important source of correlation to consider in tree breeding.

Box 5.1 Two Simple Models of Genetic Correlation

Box 5.1.1. Genetic Correlation from Linkage Disequilibrium

Assume that locus A influences trait X exclusively and locus B influences trait Y exclusively, and that both are biallelic loci with additive gene action. Further assume that both loci are in Hardy–Weinberg equilibrium. Then,

	Trait X			Trait Y		
Genotype	$A_1 A_1$	$A_1 A_2$	$A_2 A_2$	$B_1 B_1$	$B_1 B_2$	$B_2 B_2$
Frequency	p^2	$2p(1-p)$	$(1-p)^2$	q^2	$2q(1-q)$	$(1-q)^2$
Value	a	0	$-a$	b	0	$-b$
Mean	$\mu_x = a(2p-1)$			$\mu_y = b(2q-1)$		
Variance	$\sigma_x^2 = 2p(1-p)a^2$			$\sigma_y^2 = 2q(1-q)b^2$		

If we further assume that the two loci A and B are linked together on the same chromosome, then because of the linkage we need to consider both loci simultaneously. Therefore, it is necessary to deal with gamete types and frequencies, instead of allele types and frequencies, to discuss mating. When there are two loci with two alleles per locus, four gamete types are possible:

Gamete types	$A_1 B_1$	$A_1 B_2$	$A_2 B_1$	$A_2 B_2$
Gamete frequency	x_1	x_2	x_3	x_4

(Box 5.1. *Continued.*)

From the gamete frequencies the allele frequencies can be derived such that

$$p = x_1 + x_2 \quad \text{and} \quad q = x_1 + x_3$$

When the random mating is made, then 16 different genotypes can be generated and the genotype frequencies are products of corresponding gamete frequencies, assuming that the progeny population is extremely large. For example, the progeny genotype $A_1 B_1/A_2 B_1$ has the frequency of $x_1 x_3$. Of the 16 progeny genotypes, six pairs can be combined. For example, $A_1 B_2/A_1 B_1$ and $A_1 B_1/A_1 B_2$ are considered to be the same genotype. Then there are 9 genotypes and frequencies to be considered in the progeny population as follows:

Genotype	Genotypic frequency	Trait value X	Trait value Y
$A_1 B_1/A_1 B_1$	x_1^2	a	b
$A_1 B_1/A_2 B_1$	$2x_1 x_3$	0	b
$A_2 B_1/A_2 B_1$	x_3^2	$-a$	b
$A_1 B_1/A_1 B_2$	$2x_1 x_2$	a	0
$A_1 B_1/A_2 B_2$	$2(x_1 x_4 + x_2 x_3)$	0	0
$A_2 B_1/A_2 B_2$	$2x_3 x_4$	$-a$	0
$A_1 B_2/A_1 B_2$	x_2^2	a	$-b$
$A_1 B_2/A_2 B_2$	$2x_2 x_4$	0	$-b$
$A_2 B_2/A_2 B_2$	x_4^2	$-a$	$-b$

From the above table the covariance between trait X and Y can determined.

$$\text{Cov(XY)} = abx_1^2 - abx_2^2 - abx_3^2 + abx_4^2 - \mu_X \mu_Y$$

$$= ab[(x_1^2 - x_2^2 - x_3^2 + x_4^2$$

$$- (x_1 + x_2 - x_3 - x_4)(x_1 + x_3 - x_2 - x_4)]$$

$$= 2ab(x_1 x_4 - x_2 x_3) = 2ab\text{D}$$

where D $(= x_1 x_4 - x_2 x_3)$ represents the linkage disequilibrium. Therefore, the genetic correlation r between X and Y is

$$r = \text{Cov(XY)}/\sqrt{\sigma_X^2}\sqrt{\sigma_Y^2}$$

$$= \text{D}/\sqrt{p(1-p)q(1-q)}$$

From this expression, it is clear that the genetic correlation depends on the linkage disequilibrium, and that at equilibrium (i.e., D = 0) the genetic correlation is zero.

Box 5.1.2 Genetic Correlation from Pleiotropy

Assume that loci A and B jointly influence traits X and Y, and that the gene actions are both additive as follows:

Genotype	$A_1 A_1$	$A_1 A_2$	$A_2 A_2$	$B_1 B_1$	$B_1 B_2$	$B_2 B_2$
Trait X	a_X	0	$-a_X$	b_X	0	$-b_X$
Trait Y	a_Y	0	$-a_Y$	b_Y	0	$-b_Y$

(Box 5.1. *Continued.*)

If we further assume that the two loci A and B are not linked (are independent), then there is no need to deal with gamete frequencies. Assuming random mating, the genotype frequencies can be expressed using allele frequencies directly. Genotype values and frequency distribution are summarized as follows:

Index	Genotype	Genotypic frequency	Trait value X	Trait value Y
1	A_1B_1/A_1B_1	p^2q^2	$a_X + b_X$	$a_Y + b_Y$
2	A_1B_1/A_2B_1	$2p(1-p)q^2$	b_X	b_Y
3	A_2B_1/A_2B_1	$(1-p)^2q^2$	$-a_X + b_X$	$-a_Y + b_Y$
4	A_1B_1/A_1B_2	$2p^2q(1-q)$	a_X	a_Y
5	A_1B_1/A_2B_2	$4p(1-p)q(1-q)$	0	0
6	A_2B_1/A_2B_2	$2(1-p)^2q(1-q)$	$-a_X$	$-a_Y$
7	A_1B_2/A_1B_2	$p^2(1-q)^2$	$a_X - b_X$	$a_Y - b_Y$
8	A_1B_2/A_2B_2	$2p(1-p)(1-q)^2$	$-b_X$	$-b_Y$
9	A_2B_2/A_2B_2	$(1-p)^2(1-q)^2$	$-a_X - b_X$	$-a_Y - b_Y$

From this table the mean, variance, and covariance of the traits can be derived as follows:

$$\mu_X = \sum_{i=1}^{9} f_iX_i = a_X(2p-1) + b_X(2q-1)$$

$$\sigma_X^2 = \sum_{i=1}^{9} f_iX_i^2 - \mu_X^2 = 2p(1-p)a_X^2 + 2q(1-q)b_X^2$$

$$\mu_Y = \sum_{i=1}^{9} f_iY_i = a_Y(2p-1) + b_Y(2q-1)$$

$$\sigma_Y^2 = \sum_{i=1}^{9} f_iY_i^2 - \mu_Y^2 = 2p(1-p)a_Y^2 + 2q(1-q)b_Y^2$$

$$\text{Cov}(XY) = \sum_{i=1}^{9} f_iX_iY_i - \mu_X\mu_Y = 2p(1-p)a_Xa_Y + 2q(1-q)b_Xb_Y$$

where

f_i represents i^{th} genotype frequency,

X_i represents i^{th} X-trait value, and

Y_i represents i^{th} Y-trait value.

Therefore the genetic correlation

$$r = \text{Cov}(XY)/\sqrt{\sigma_X^2}\sqrt{\sigma_Y^2}$$

$$= \frac{[2p(1-p)a_Xa_Y + 2q(1-q)b_Xb_Y]}{\sqrt{[2p(1-p)a_X^2 + 2q(1-q)b_X^2]}\sqrt{[2p(1-p)a_Y^2 + 2q(1-q)b_Y^2]}} \qquad (5.4)$$

Although linkage implies the situation in which two loci are linked on the same chromosome, effects similar to linkage disequilibrium occur even when the two loci are independent. This happens under inbreeding (Weir and Cockerham, 1974). This can also occur in hybrid populations in which the allele frequencies are different in the parental populations, even if no linkage exists. With linkage, this linkage disequilibrium is expected to dissipate slowly, and if simultaneous selection is applied, a permanent disequilibrium can exist. However, these associations can change and hence the genetic covariance can change if it results at least partly from these disequilibria. The disequilibria can be maintained by selecting for the correlated traits, but in every generation, there is a decay rate that must be countered. To maintain a genetic correlation of large magnitude in the absence of pleiotropy, a high level of inbreeding and tight linkage is necessary (Lande, 1984).

An interesting effect of linkage is in hitchhiking. This is represented by a selectively neutral locus linked to the selected locus. Because of linkage, the frequency of selectively neutral alleles moves in the direction determined by the sign of the disequilibrium. This model is interesting in juvenile selection because the selection is made in the juvenile stage but not in mature stages. Therefore, if there are loci that only influence mature traits and are selectively neutral at juvenile stages, then the allele frequency of these loci may change through hitchhiking. In two-locus hitchhiking (one selected locus and one neutral locus), the sign of disequilibrium does not change with repeated selection (Asmussen and Clegg, 1981). However, when there are selected loci and neutral hitchhiking loci, then linkage disequilibrium between the two neutral loci can change sign (Thompson, 1977).

The hitchhiking effect can also occur under inbreeding. The rate at which the neutral allele frequencies move tends to be faster with inbreeding than with linkage (Hedrick, 1980). The overall behavior of the inbreeding hitchhiking model is similar to that of linked hitchhiking, including sign reversal between the two neutral loci. Generally speaking, to maintain a correlation between two traits under separate genetic control, tight linkage and heavy inbreeding are necessary (Lande, 1984).

Although these theoretical findings do not offer concrete recipes for developing a breeding program designed to maintain a high genetic correlation, breeders can conclude that inbreeding is an important tool to consider in long-term breeding because there are few other means of affecting linkage among different loci. It is also useful to keep in mind that in a large randomly mating population with no linkage between loci influencing different characters, genetic correlations are soon dissipated (Lande, 1984).

If a pleiotropic effect is expressed by a single locus with two alleles, then genetic correlation between the two traits is either 1, 0, or -1, depending on the manner the locus influences the two traits. These values are fixed, unless the gene action itself changes over generations. In more general models, such as shown in Box 5.1.1, the genetic correlation can take other values between -1 and $+1$, and can change over generations. Although decay such as found in linkage disequilibrium is not present, the genetic covariance changes as the allele frequencies of the involved loci change. For example, Fig. 5.1 describes the dynamics of genetic correlation as gene frequency changes. The figure is drawn based on a specific case of Equation 5.4 (in Box 5.1) that assumes that $a_X = a_Y = b_X = -b_Y$. This assumption means that the magnitudes of the gene action in the two loci are the same, and that Locus A influences traits X and Y in the same direction, but Locus B influences the two traits in the opposite direction. Then the expression for r is simplified such that

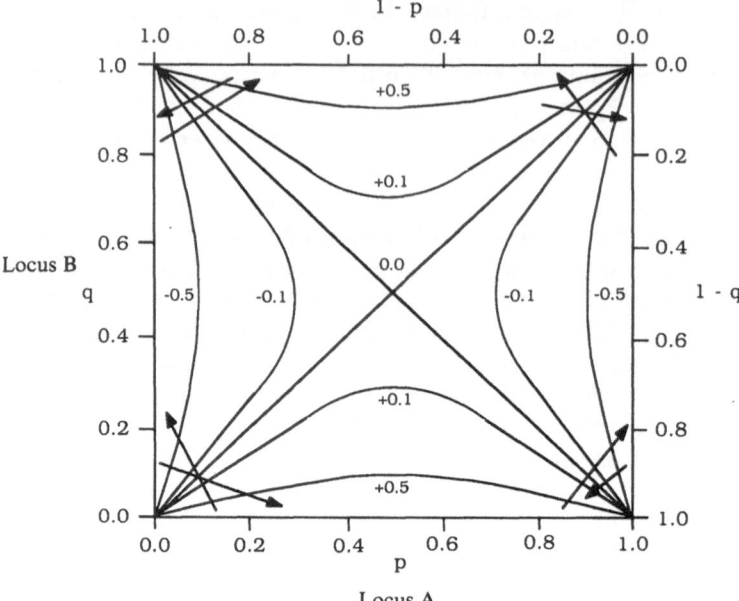

Fig. 5.1. Genetic correlation as a function of gene frequencies at two loci. (Modified from Mitchell-Olds and Rutledge, 1986, American Naturalist, 127, p. 389. © 1986 by The University of Chicago Press. All rights reserved.)

r is a function of gene frequency of the two loci only:

$$r = [p(1 - p) - q(1 - q)]/[p(1 - p) + q(1 - q)]$$

Although the assumption is unrealistic, and the model is simple, it shows some interesting dynamics for the genetic correlation that originates from pleiotropy. First, unlike the case with linkage, zero genetic correlation does not imply a lack of causal relation between the two traits. The two diagonal lines ($p = q$ or $p = 1 - q$) indicate zero correlation, and these lines do not represent any equilibrium points (because of the additive gene action). Any slight move in the allele frequencies will produce a genetic correlation. Second, as the allele frequency changes, the sign of genetic correlation can easily change.

In this particular model, it can be seen that there is a broad zone of low genetic correlation and slow change in genetic correlation in the middle gene frequencies, where progress from selection is likely to be most rapid. On the other hand, at extreme gene frequencies correlations can change rapidly, and can easily change sign (note *arrows*). This implies that at the beginning of long-term breeding or near the limits to selection, genetic correlations are likely to change rapidly, whereas the progress from selection is slow. The extension of Equation 5.4 to many additive loci influencing traits X and Y is straightforward:

$$r = \left[2 \sum_{i=1}^{n} p_i(1 - p_i)a_{X_i}a_{Y_i} \right] \bigg/ \sqrt{2p_i(1 - p_i)a_{X_i}^2} \sqrt{2p_i(1 - p_i)a_{Y_i}^2} \qquad (5.5)$$

where i indicates the ith locus.

From Equation 5.5, similar inferences can be drawn for the more generalized pleiotropic model. First, the genetic correlation is a function of gene frequencies of the loci involved. Second, zero genetic correlation does not imply an absence of causal relations. The denominator of Equation 5.5 is always positive, and the sign of the genetic correlation depends on the numerator only. Because the numerator is a sum of cross products $[p_i(1 - p_i)a_{X_i}a_{Y_i}]$, depending on the relative magnitude of the positive and negative cross products, it is possible to find a combination in which the numerator is zero. However, we can also conclude that zero genetic correlation is probably rare and transient under the additive pleiotropy model.

Although not apparent from Equation 5.5, it is not difficult to derive that the genetic correlation from multilocus pleiotropy can be maintained through selection. Cheverud (1982, 1984) suggested that stabilizing selection can be used to manipulate the joint distribution of the additive genetic values, and that stabilizing selection on the genetic correlation can be independent of the directional selection on the bivariate mean. Therefore it is possible to move the mean while preserving the genetic correlation. Because this stabilizing selection requires information from both traits, the application of such selection to juvenile–mature correlation may not have great appeal in terms of enhancing the rate of generation turnover. However, development and maintenance of such high correlations in long-term breeding could serve as the basis for deriving short-term breeding stock with high juvenile–mature correlation. Furthermore, with pleiotropy, it is possible that if high correlation existed between juvenile and mature traits, strong directional selection is more likely to destroy the correlation structure than mild selection, which is generally suggested for long-term breeding (Dempster, 1955; Robertson, 1960). The maintenance of correlation applies to other traits as well, and this information is also useful in developing multiple-trait selection strategies.

5.4 Multitrait Selection Techniques

Whether intended or not, the selection of trees for one trait will inevitably result in changes in other traits. Sometimes this is an accident of sampling, but often is the result of the real correlations that exist between the selected and other traits. If a zero correlation exists, then any changes in related traits are random and can only be controlled by reducing the sampling effects of randomness by such means as maintaining large selection populations. However, if correlations exist, then correlated responses must be expected and the breeder may choose to influence them genetically, selecting trees on the basis of their performance in the several traits. Quite often, the breeder in fact wishes to change several traits simultaneously, and hence will select on the basis of several trait evaluations. In these cases, it is usually assumed (but as we shall see, not necessarily true) that the multiple traits can be incorporated into a single function that defines value. Then, the breeder has only to determine how strongly to select on each of the traits and the method of selection. The three classical means of doing this are called *tandem selection*, *independent culling level*, and *index selection*. These have been described in more detail elsewhere (e.g., Namkoong, 1979), but can be summarized briefly here.

Tandem selection involves setting truncation levels for each trait, but would do so in sequential generations, usually one trait per generation, and hence would be applicable to organisms in which breeding is rapid enough that the gains in the component traits are made in a timely manner. This might also occur when the economic importance of traits is not immediately apparent and generations of breeding populations are sequentially subjected to new selection objectives. In such cases, the effective population size may change between episodes of selection, and predicting gain becomes very difficult.

In fact, over the generations of a recurrent selection breeding system, we could consider the set of traits emphasized for selection in one generation to be only partially dependent on the selection scheme of the previous generation. In many forest tree breeding programs, the needs for trait improvement change from generation to generation, and the ability to measure and test for trait responses may also change. In some situations, it might be important to select only for survival and establishment on harsh sites in the first generation. If those traits are sufficiently improved as general forestry practices become more established during a first generation of tree improvement, then emphasis may shift to volume growth, performance on other planting sites, or other traits. By the time of the second and later generations, presumably much more information on developmental genetics and physiology would permit more refined testing and selection for particular growth response types. Hence, the heritability of traits can be expected to increase in general.

If the set of traits for improvement are previously determined, and a single, final breeding population is to be generated, the relative culling levels in each generation should be set to give the maximum possible economic benefit for a given population size. Assuming that the size of the breeding population is determined from considerations of minimum N_e, and the size of the population from which to select is large and determined, the total selection proportion is then fixed. The problem then is to allocate the selection, and because this is time dependent, it would clearly be far more efficient to postpone these phases of tandem selection to those generations when selection is more feasible.

An analogous sequence of selection objectives was suggested by J. Wright (1976) for the case of provenance selection to be followed by family or individual selection in a second generation, if it proved feasible. In the Republic of Korea, following the deforestations during World War II and the Korean War (1950–1953), the first requirements of forestry were stand establishment on denuded hills to provide cover and improve soil and water-holding capacities, and to supply a minimum amount of firewood. Subsequent programs were then focused on growth and yield and the traits needed for the different soil, climate, and market demands.

In contrast, independent culling level (ICL) sets truncation levels for each trait, and only those trees that possess trait measures above the truncation points in all traits would be included in the selected breeding population. If the traits are uncorrelated, then the gain in each trait is independent, and the gain in value is determined by relative economic weights and the expected independent gain in each trait. Because the gain in each trait is a function of its heritability and the selection differential, it can then be determined which allocation of selection proportions maximizes economic gain. However, if the traits are correlated, then the optimum allocation of selection proportions is more complicated to analyze, because selection on one trait affects the response in others.

For many practical reasons, ICL is used during the course of a generation of breeding activities. For example, whenever testing or measurement expenses are incurred, such as in progeny testing, testing for general combining ability, or testing for special wood quality or resistance characteristics, it is useful to cull individuals beforehand from large candidate populations. It is also feasible to cull candidate trees on the basis of traits that are expressed early in the life history and to leave those traits which can only be selected at later ages for final culling. It is therefore useful to not only consider the benefits to derive from ICL selection, but also the cost savings associated with sequential culling within a single breeding cycle. Periodic culling can then be more economically efficient than index selection. Otherwise, in terms of genetic gain alone, index selection will always provide greater expected gain for a given total selection differential than either tandem selection or ICL.

Selection indices generally assume that correlations exist among traits, and because the same index method is applied to all individuals, there is no question of allocating the selection differential. Instead, the problem is stated in terms of how to give relative weightings to individual traits such that the population advances in the combined trait direction of maximal economic gain. We create the index in a linear form: $I = b_1 y_1 + b_2 y_2 + \cdots$, assuming that each phenotypic trait measure is related to value (V), which itself is a linear function:

$$V = a_1 g_1 + a_2 g_2 + \cdots$$

where g_i is the inherited or genotypic value of the entries, and a is the economic weight.

We maximize gain in V by maximizing the correlation between I and V. By least squares we derive:

$$\begin{bmatrix} b_1 \\ b_2 \\ b_3 \\ \vdots \end{bmatrix} = \begin{bmatrix} \text{Cov}(X_1, V)/\sigma_{X1}^2 \\ \text{Cov}(X_2, V)/\sigma_{X2}^2 \\ \text{Cov}(X_3, V)/\sigma_{X3}^2 \\ \vdots \end{bmatrix}$$

$$= (G)P^{-1}a$$

where

G is the genetic covariance matrix
P^{-1} is the inverse of the phenotypic covariance matrix, and
a is the economic weight given to the component traits to form the economic value function.

Obviously, the error of estimate for the b_i depends on how well G and P^{-1} are estimated, and for large numbers of traits, these estimates can be very poor. Williams (1962) and Patel et al. (1962, 1969) in fact suggested that the errors can be so large as to rank index selection nearly random with respect to value. Hence, with large error, merely using the a_i weights, or weighting them by trait heritability alone, can be almost as good as index selection. Nevertheless, there are modified and restricted index procedures and better estimating methods currently available such that some index procedures can be very useful (Baker, 1986).

A useful geometric translation of index selection is to consider a space in the p-trait dimensions with one axis for each of the traits. The individual entries may then be located by their value in each trait and the b coefficients are merely the direction cosines of the index line (Namkoong, 1976).

5.5 General Strategies for Multiple Breeding Objectives

Beyond the statistical estimation problems, and beyond the selection and breeding operational problems, there is a more fundamental problem of choosing the economic objectives of tree breeding. There is an inherent uncertainty about what the breeding goals ought to be, not only about which traits to improve but also about the value of increasing or decreasing measures such as wood specific quality, and the relative value of that versus simply increasing dry matter production. Quite distinct from the problem of uncertainty is the existence of multiple objectives for forest management and the role of tree breeding in providing managerial options. Even in the absence of any uncertainty about the relative economic values, there are often different ecological and economic goals within a forest region and hence different breeding objectives among forest compartments. Further, because the ecological and economic factors that govern forest management often shift from generation to generation, there is a temporal variability in breeding objectives.

5.5.1 Incorporating Multiple Objectives in a Single Population

If tree breeders wish to be economically effective, some of these variations can be accommodated by breeding for traits that are of uniformly high value regardless of economic or ecological condition, or alternatively, by breeding multiple populations with different sets of trait values. The first option is the simplest. Under stable conditions, the average relative values of traits may be well known and estimable with little error. In some agricultural crops, this situation may exist, at least for long enough periods of time relative to the generation length, that average values do not vary much. In such cases, simply breeding to maximize the average values is reasonably easy to justify economically.

Such degrees of certainty may be quite rare in forestry, however, and hence breeding for perhaps unlikely value functions can be highly inefficient. While an average condition is possibly more likely than any other foreseeable event, the greater the uncertainty, the greater the risk that the breeding effort is misdirected. In such cases, the optimum strategy may not be to maximize an expected average gain, because it may be known only with high error of estimate. If the uncertainty were such that the error on predicted values was narrowly distributed around a mean, then one might wish to treat the case as a deterministic one. In what might be the more common situation, however, a relatively high variance on the value function exists, and an estimated average function may have too high an error to attempt a definition of optimum breeding direction cosines. An alternative strategy is to determine simultaneous confidence limits within which the optimal direction exists at given probabilities. Breeding evaluations may then be made on the assumption that a maximum error and an associated minimal gain can be made for a given range in direction cosines. Alternatively, a breeder may wish to minimize the probability of certain errors occurring, such as some limit on misdirection, and may instead choose direction cosines with minimal gain objectives.

Under conditions of high uncertainty, such that the error distribution on the predicted value function is too large to reasonably derive an estimated mean with any reasonable probability of accuracy, the concept of maximizing expected value may have to be totally discarded. In such cases, it may be possible to describe several value functions for the combined traits, no one of which is any more likely than the rest to reflect the actual value function at harvest time. Then, maximizing a minimum expected gain may be a more reasonable, if conservative, strategy to follow, because it would guarantee that certain minimal values can be achieved regardless of which value function actually exists at harvest. The optimum trait direction for the breeder to follow is thus defined as that which imparts the highest minimal gain that can be achieved. As an oversimplified example, consider only one trait, wood specific gravity, which is highly correlated with cell wall thickness, and assume that one of four situations illustrated in Fig. 5.2 may occur: (1) Thin-walled fibers become very valuable and wood of low specific gravity has high value. (2) Thin-walled fibers are of some value but the loss in fiber yield of low specific gravity wood almost offsets the value of the wood quality gain. (3) Cell wall thickness does not affect value and the increased fiber yield of high specific gravity wood increases its value. (4) Thick-walled fibers are of some value and accentuate the increased value of high specific gravity wood.

If these four adequately define all value functions, the ∗ marks in Fig. 5.2 represent the extreme points of the possible solution set. The value of specific gravity that maximizes the minimum gain is indicated by a circle and would be the optimum

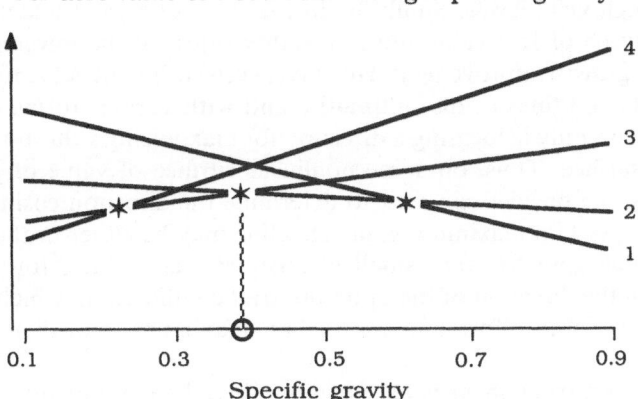

Per-acre value of wood with average specific gravity

Fig. 5.2. Maximin problem for values of specific gravity in four possible situations: (1) strongly decreasing value; (2) moderately decreasing value; (3) moderately increasing value; and (4) strongly increasing value.

point toward which the population should move. It can be seen that discontinuities in the value functions would cause a problem in arriving at a solution.

In any real situation, the economic functions would be more complicated and involve more variables, but they could be easily solved, since the solution of maximin games can be found by standard linear programming (Hadley, 1964). Difficulties with continuous, nonlinearizable functions remain (Owen, 1968), but with modern computers, good linear approximations for small intervals can be made, the expanded set of restrictions can easily be handled, and all problems of discontinuities and boundary values have been eliminated. Therefore, solutions for such linearizable functions in these essentially two-person, zero-sum games can always be found and an optimum point or direction defined. Therefore, a direction for moving the breeding population can be defined even in these situations.

These programs for maximizing a known value function generally assume linear value functions for the traits involved. This is obviously rarely true in forestry; several forms of nonlinearities are fairly common in forestry that cannot be linearized by logarithmic or polynomial transformations. Slightly more difficult to handle are cases in which discontinuities exist in the relationship between physical measures and value, such as between stem diameter and stem value when it jumps from pulp size to pole, saw-log, and veneer-log sizes. In addition, the degree of pest resistance may exhibit a relatively flat value function until some minimal levels are reached, after which a linear function may exist until high resistance levels are reached, when further increments in resistance add little to final crop value, especially if some natural thinning is expected. In many such cases, approximate value functions may be assigned and linearized, even when multiple discontinuities exist.

Only slightly more complicated and difficult are those cases in which the value of one trait depends to some extent on the value of other traits, when trait values are interdependent. The joint value function then requires some iterative evaluation as, for example, when both volume growth and wood quality are interdependent and both depend on survival and pest resistance levels. For example, pest resistance may be of relatively little value at low growth levels but can assume an exponential value function at high yield levels. Similarly, increasing wood yield under high risk of mortality may be of low value until mortality rates can be lowered to warrant investment in growth improvement. However, even such joint value functions, with various points and lines of discontinuities and with various forms of curvature, present problems only in locating a direction for maximizing value on a non-linear, but known, surface. Thus, on some nonlinear surface of value of possible trait combinations, the breeder can seek to determine the direction cosines of the line which would give him maximum gain. This line may be either in the direction of maximum value gain for very small changes in trait values, for example, the gradient, or in the direction of the optimum trait combination, which may require a long-term breeding effort and more than maximum immediate gains could promise.

In the case of a truly linear value function, these lines are identical, and the index coefficients that are derived by traditional methods are the direction numbers of the planes of equal value, which are perpendicular to the gradient. Determinations of the direction numbers of planes of equal value or the associated direction cosines of the line normal to those planes are essentially similar operations, as one can be determined from the other. Therefore, other than change the direction of maximizing value gain according to the present mean value of the population, the foregoing nonlinearities cause no theoretical problems in breeding in whatever direction is determined to be optimum.

5.5.2 Incorporating Multiple Objectives in Multiple Populations

All of the foregoing breeding approaches are for single-population breeding in which the attempt is to maximize a probable or, at least, a minimum value. An alternative is for the breeder to offer the forest manager different kinds of trees, which possess different trait–value combinations. That is, it is possible to establish a system of multiple population varieties in which each population is developed for different potential uses and which together form an array of populations for a variety of possible future needs. Genetic management then involves not only increasing trait performance, but also involves choosing the number and distribution of breeding populations and the means used to develop the population array.

The concept of using multiple-population replicates in breeding is not new (Baker and Curnow, 1959). In a set of small populations, compared to a single large one, the genetic gain from breeding will not be the same among them because of sampling variations and other uncontrolled errors. The average of the several populations will not come up to the average of the single large population, even if the same overall selection intensity were kept, because of the smaller effective population size in each. However, two advantages accrue: (1) the variation among the subpopulations can be expected to be large enough for some of them to exceed the average gain of the large single population, and (2) the different subpopulations, even if the same mean were obtained, will be using and fixing different alleles. Therefore, a selection among the subpopulations in future breeding cycles, and crossing among them, can be expected to give greater gains and to regenerate any lost useful variations in recombined populations.

In a wider sense, the directions of selection in multiple traits can be more purposefully established for a range of economic or silvicultural objectives. It can be shown that, if there is any variation in future requirements, multiple-population breeding for an array of populations around the mean breeding objective gives a higher average gain than single-population breeding for the mean expected requirements. Thus two or more subpopulations can be directed to somewhat divergent combinations of traits or adaptabilities and the one which happens to be closer to the realized future condition can then be used for commercial seed production. At that time, the several other populations not necessarily used there would be reevaluated for further breeding to again reach an optimum array for the subsequent generation. It is possible that several such populations would be similarly directed and be merely replicated populations in the manner of Baker and Curnow (1959). It is also possible that population hybridization would be chosen for immediate use or for future breeding if the new directions required such a genetic merger.

Thus, if breeding can be done with 100 trees, there would be no or little extra work to breed in 10 units of 10 trees each, so long as particular sets can be rapidly expanded to provide the commercial seed required. As much or as little extra work can be done on the subpopulation as on the large single population. In fact, for some of the subpopulations, more intensive development may be pursued than in others if those populations justify more attention. If some sets are directed toward different site adaptabilities, then the separate subpopulations may be used in specific zones.

Given varietal populations with different trait combinations, the forest manager can simultaneously use different varieties in mixes within stands or establish separate varietal stand blocks, or use sequences of different varieties singly or in mixes, as desired. With this broader concept of multiple-population breeding, the concept of metabreeding (Namkoong et al., 1980) is fitting. The species or metabreed is conceived of as a broad mix of multiple, independent populations, possibly bred for some common objectives, but in general bred for different trait objectives. Some of these populations may have identical objectives and serve as in Baker and Curnow's replicate-population breeding system. For some populations, however, more extreme trait combinations may be sought to develop trees that are suitable for other than "average" purposes. Such populations may serve directly for seed production, or may serve as a source of genes for hybrid or combination breeding. As such, they may be more useful if they are selected for even more widely divergent trait measures. For these purposes, we require some understanding of the form of gene action.

While the concept of metabreeds seems to involve substantially more breeding effort than the simple, single-population systems, it will be seen that the total effort is marginally greater. By placing emphasis on rational planning, it does make the planning of breeding programs far more rigorous. The long breeding cycles and rotation ages of most forest trees require such planning to capture the possibilities of recurrent selection. The luxuries of using annual breeding and harvest cycles to rapidly generate new varieties are generally not available, and hence adjustments of populations for new selection goals require the establishment of an array of populations, some of which might be useful in any of several possible futures. Thus, more than in most crops, tree breeding requires anticipatory breeding and can ill afford post facto reactions to breeding opportunities. If a few hundred trees are to be used in breeding and are needed for future gains, but only a few tens are needed in any breeding population, then instead of using all in a single breeding population it takes no more work to organize breeding into multiple populations. Future adjustments in a multiple tandem selection program in each of the populations can then be made to provide further diversification or to better track the changing needs of forests.

5.6 Combining Gene Actions, Breeding Objectives, and Breeding Techniques

We discussed breeding methods that are designed to deal with nonadditive or mixed gene actions for a single trait (Chapter 4). If it is necessary to breed for multiple traits, and if nonadditive gene actions are important for some of the traits, then breeders will have to combine mixed breeding methods, which were discussed in

Chapter 4, and the multiple-trait selection techniques discussed in this chapter. Such joint breeding action is likely to increase the problems associated with design and test of experiments. In this section we discuss the joint breeding action and associated design and testing problems in general terms. Any detailed discussion requires defining a specific breeding program, and is beyond the scope of this book.

5.6.1 Multiple Gene Actions and Multiple Traits

A simple means of combining multiple gene actions and multiple traits is to ignore dominance gene actions and to breed two populations using SRS for long-term breeding,with the expectation that such nonadditive gene effects will remain in the populations. Within each population any of the three multiple-trait selection techniques may be used. For short-term breeding, the two populations may be crossed to form a hybrid population. Alternatively, a RRS may be used, selecting for all traits with additive and nonadditive gene actions where multiple-trait selection technique is applied to the hybrid progeny population.

If a SRS is chosen for one trait (or one set of traits), but a hybrid system for a second trait, then the parents for the hybrid intercrossing can be chosen in a system of tandem selection, or ICL. Although SRS is deployed to select some traits, a minimum of two breeding populations and one progeny population are involved in the complete cycle of selection, because hybrid intercrossing is involved in the selection process. As is the case with any tandem selection, the number of breeding generations involved in the complete cycle of selection equals the number of traits or sets of traits selected in tandem. Figure 5.3 shows a two-trait tandem selection procedure with mixed breeding. In this procedure, within-population selection is used for the first trait, and then the second trait is selected on the basis of a hybrid progeny test. When more than two traits are involved, a modified tandem selection may be used. In this scheme, sets of traits instead of single traits are selected in selections 1 and 2 of Fig. 5.3. In each stage of selection, either ICL or index selection may be used to select traits that belong to the set. In ICL, regardless of the number of the traits involved, one breeding cycle with a two-stage selection will complete the cycle of selection (Fig. 5.4). This procedure is similar to the Modified RRS described in Chapter 4. The order of within-population selection and selection based on

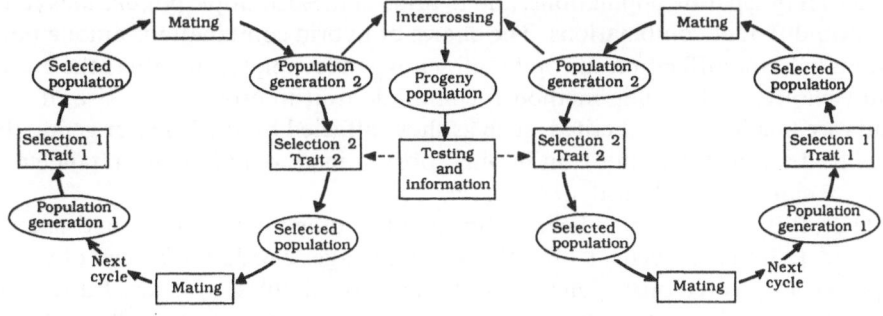

Fig. 5.3. Two-trait tandem selection with mixed breeding.

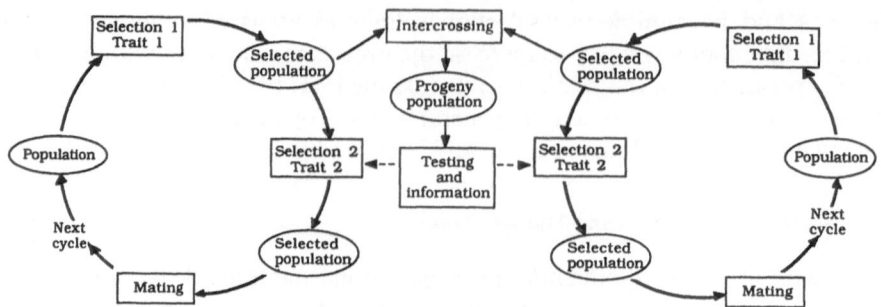

Fig. 5.4. Multiple-trait independent culling-level selection with mixed breeding.

hybrid progeny test may be reversed, but may hold no theoretical advantage while probably costing more to operate.

It is also possible to incorporate a form of index selection in which the entries are judged as potential parents on the basis of indexed values of hybrid progenies; the index weights are estimated with the variances and covariances of the hybrid progeny population. Thus, these combined methods are a straightforward application of the multitrait selection techniques, and the advantages and disadvantages of each selection system remain. Unfortunately, there is no theoretical foundation to evaluate different combined breeding methods. The two methods discussed above simply represent two examples that might be operationally feasible in tree breeding.

The existence of multiple-trait breeding objectives, however, does influence at least two other aspects of breeding. The effect of linkage among traits may be viewed as being no different for multiple traits than for single traits, because breeding for a single value function of multiple traits is only taking a one-dimensional transformation of those traits and hence is merely another complicated trait. However, the effects of linkage, its dissipation over breeding cycles, and its effects on epistatic effects are less likely to be linear and thus changes in linkage are likely to have strong effects on the ease of single-population breeding.

Consequently, the second effect is the increased importance of multiple-population selection methods, because more opportunities exist to breed for different alleles in separate populations, and hybrids of these multiple populations can be made in different combinations. The choice of hybrid combinations among populations bred for different trait combinations expands rapidly, and offers the breeder many choices of breeding method for multiple-trait improvement. Also, if some traits are qualitatively inherited, such as those affected by single major genes, then the breeder can choose this trait in only some or in all populations, in tandem, by ICL, or simultaneously with quantitative traits.

Thus, it is possible, by judicious choice of breeding method, trait, and selection method, that the breeder can offer the forest manager a wide assortment of varietal types at no greater operational expense than the simplest systems. Further, the difficulties of using index selection methods can be confined to a small set of traits.

Even if it is possible to consider selecting multiple populations using different indices in each (Namkoong, 1976) one can use index methods on some traits and use ICL or tandem on most.

5.6.2 Allocation of Resources and Test Design

The results of breeders' efforts will thus become far more useful to forest managers as they move beyond the simple selection and seed orchard management problems of the first generation. However, breeders will then be constantly faced with the challenge of optimally allocating materials to estimate several parameters, and for this a well-defined combined breeding method will help make judicious decisions on resource allocation. Clearly, a major objective of test designs will be to reduce errors of estimate of family or individual genotypic values, and to estimate variances and covariances of environmental as well as genetic sources of variance so that a good allocation of effort among traits and stages of selection can be determined (Section 5.2). At the same time, many breeders will use selected parts of the test materials themselves to be the parents of the next breeding population. We have so far approached each of these three objectives as independent problems, and especially the last one as merely being a function of allocating the selection differential. We have also assumed that production of the seed or clonal material for actual forest regeneration programs are separately managed, as in seed orchards.

Some species may be sufficiently important for forest management that these three functions can be separately managed. But for many species, particularly those of secondary importance in an artificial regeneration system, at least two of the functions must be performed by a single set of materials. If multiple-population breeding (Section 4.3) is used, it may be possible to carry along recurrent selection breeding populations separately from a smaller set of estimation populations (index populations) with or without testing family means, as Kang (1980) suggested for jack pine breeding in the Lake States. However, it might be necessary to consider how a single set of materials could be managed that could be used simultaneously for testing and estimation, at the same time that some of the materials it generates can be used for breeding.

There are obvious conflicts in the optimum allocation of materials for these objectives. Small variances on family mean estimates requires large numbers of progeny per family, while small variances on genetic variance component estimates requires large numbers of families. In one sense, we consider the genetic variation among individual contributors to the gene pool to be a random source of variability, but in another sense, the breeding values of specific individuals in one generation of breeding are fixed sources of variance.

Nevertheless, in many cases once the breeding population is closed and only intrafamily selection is followed, all the criteria of goodness become functions of the design parameters and allocation proportions. Thus, at least theoretically it is possible to search among all possible allocations of materials to determine their quality with respect to each objective for given ranges of the parameters. It would then be a simple matter to determine which value function to apply to the satisfaction of the three objectives to determine a best or a least unsatisfactory solution. This has not yet been done, nor is this the place to discuss methods of accomplishing such a task in general. Lacking a general solution, however, some particular features can still be deduced.

In terms of the genetic design, both the testing and the estimation objectives for general combining ability do not require many crosses per parental entry to achieve low error variances. Because the total number of entries can be relatively small per population, and the total breeding population need not be more than a few hundred, a few crosses per entry constitutes a large but manageable set of crosses. As noted by Namkoong and Roberds (1974), the differences between mating designs for testing purposes is not large so long as the number of crosses is greater than four, but the efficiencies for estimation differ considerably. For breeding purposes, the factorial or tester systems of mating most severely reduce the inbreeding effective population size. If family selection is to be practiced, then the largest number of intercrosses in a diallel type of mating pattern will give the highest probability of having made crosses of parents with the best general and specific combining abilities.

The marginal gain declines, however, as the number of crosses gets very large while the available resource for breeding remain constant. Therefore, something less than complete intercrossing is necessary for cost effectiveness. As generations advance and less family selection becomes possible, the number of crosses per entry desired for breeding purposes will decline and design optimality will be dictated more by estimation and testing objectives. For these purposes, partial diallel designs provide many options for efficiency. In fact, mixed designs might prove more operationally feasible, with one set of diallel crosses made for estimation and supplemented by a set of partially balanced tester parent crosses for family mean error reduction.

With unbalanced field designs, use of unbalanced mating designs may be either balanced (i.e., complete) across replicates, or some confounding of imbalance may be useful. For testing purposes, a balance of the crosses among the unbalanced replicate field plots may be necessary, but for estimation, some confounding could be of practical use in the field to reduce replicate size while not substantially reducing estimation efficiency.

Because most forestry operations take several years to complete, it may also be useful to be able to phase operations over several years. There often is no necessity for the three objectives to be synchronously satisfied, and hence some operational options may exist to first satisfy the testing objectives, for example, and then the breeding and estimation objectives with supplemental crosses and plantings in later years. Alternatively, one governmental or cooperative organization may develop the crosses and plantings for one set of objectives, while industry breeders may use the same materials in different designs for other objectives.

CHAPTER 6

Environments

In Chapters 3 through 5 we discussed breeding systems with incremental complexity. Throughout the discussion we concentrated on the genetic aspects of breeding and made little reference to environments. For example, in addressing the concept of heritability, we used the genetic variance as a measure of the relative importance that genotypic differences have for an assumed set of environments in generating a total phenotypic variance. Although it is not the intention of this chapter, we may reverse the thought process and conceive of environmental differences as being in part an agent in generating that total phenotypic variance for an assumed set of genotypes. Although we pay greater attention to environments, our ultimate goal in this chapter is to develop a holistic breeding concept that accounts for both genetic and environmental sources of variation.

In Section 6.1 we extend the foregoing thoughts on the joint effects of genotypes and environments in producing phenotypes. In Section 6.2 we discuss patterns of response to environmental variation. In the past, breeders used linear approximations to represent plant response to environmental gradients. This approach was convenient and may have been useful for some crop breeding, but is unrealistic for forest tree breeding. Therefore the need for developing measures for nonlinear response is introduced. In Section 6.3 we discuss the need to determine multiple optimum combinations of genotypes and environments. Environmental zonation is introduced as a means of finding such optima, and means of combining multiple-population breeding (discussed in Chapters 4 and 5) with environmental zoning is presented. In Section 6.4 competition is discussed as a special form of environment with which breeders must deal.

6.1 Genetic, Environmental, and Phenotypic Expression

Because we cannot measure a genotypic effect without an environment, and environmental effects are expressed only through genotypic responses, we cannot measure one without defining the other. Therefore, it is not meaningful, in general, to suggest that genotypically caused variance and an environmentally caused variance are independent factors that separately generate phenotypes. It is not accurate to describe a fixed heritability or to try to estimate one without defining a fixed set or a range of both genotypes and environments. This is particularly critical when the environments referred to are not well defined or when they may change. In forestry, where the physical environment is less well controlled than in agriculture, and where silvicultural regimens are changing, it is therefore difficult to define "true" genetic effects and variances. Yet testing is conducted to estimate a genotypic mean as if that is fixed and as if a "true" environment exists that would allow the genes to be properly expressed. Therefore, instead of trying to capture some nonexistent fixed effects, we wish to discuss both genetic and environmental effects within the context of a fixed set or sample from some domain of both.

We often refer to a set of environments within which genotypic expressions differ as a random variable, the mean of which must be the genotypic mean. These environments may be composed of many distinct effects that we may not be able to separate, such as soil and atmospheric, moisture, temperature, and pH, which affect traits of interest. It may include the tree's own physiological state, and other organisms that also affect trait expression. Some of these factors may interact so complexly and vary so rapidly over time, and over such small distances, that we can either fail to recognize distinct factors or cannot manage them on so fine a scale. For practical purposes, such variations become part of an irreducible noise or error variance. However, some factors may be sufficiently strong and direct in effect, and vary so slowly, that we can recognize, plant, and manage different populations in zones or possibly even manage that environmental variation. In these latter cases, then, the geneticist and silviculturist can affect each other's prescriptions, and it becomes useful to examine the environmental effects on a set of genotypes and vice versa.

As discussed in Chapter 5, some efficient experimental designs can be used for wide-scale estimation of response functions for many genotypes. However, relatively few such experimental results have been reported for forest trees. In one test of loblolly pine families, Roberds et al. (1976) observed that the pattern of height growth response to increases in urea application was curvilinear during the first season after application.

In natural forests, of course, it is clear that genotypes exhibit variation in environmental response functions, as observed in most provenance tests. The response form for height growth in Ponderosa pine clearly shows a peaked functional form with elevation (Namkoong and Conkle, 1976).

Under these conditions, in which the response of genotypes to an environmental span differs, the relationship between heritability and gain breaks down because the relative genotypic values can change and because the genotypic variance may depend on the environments included in the population of inference. Thus, the value of breeding will depend on which set of environments is inferred and how that variation may be managed.

Finally, it is also clear that phenotypes have no value outside an economic system of evaluation. If the economic environment changes slowly relative to the management cycle, then a constant evaluation system can be assumed. However, if economic value changes over the geographic span of a forestry operation, or during a few generations, then the value of genotypes depends on which particular economic conditions prevail. These economic variations may be related to environmental variations, such as when the value of disease resistance varies with environmental variations in disease prevalence. Even if independent, however, the value of genotypic performances may be significantly affected by the economic environment, and its scale of change can affect genotypic value performance and ranking. We discuss these effects in Section 6.3.

6.2 Patterns of Response to Environmental Variation

Perhaps the earliest demonstrations of environmental response functions in forest trees were the provenance tests laid out by Langlet (1963) for *Picea abies* over a latitudinal gradient. For several of the southeastern U.S. pine species, the Southwide Pine Seed Source Studies, initiated by Wakeley and later by Wells and Wakeley (1966), also provided substantial data on how whole populations or families displayed regular variations in growth response depending on the location of the planting.

In this chapter, we are primarily concerned with the form of the response displayed by particular genotypes, but this is often confounded by interest in the origins of such response variations. In Chapter 8, we consider the sources of such variations, but we cannot completely divorce our measures of the importance of response variations from the sources of that variation. In particular, if local populations differ in their adaptability to soil variables, it would be surprising if their progeny do not display some performance variations over differences in soil types.

It might be expected that adaptational clinal variations of populations over an environmental gradient would contain genotypes that vary gradually in their responses to a gradient in that environmental variable. However, the effects of selection and mating pattern are confounded in genotypes found in natural populations, and the observations of clinal or ecotypic patterns of variation in natural populations is not necessarily indicative of the kinds of response functions that individual genotypes can display. To sample response functions adequately we require wide population samples, especially if special site adaptabilities are sought.

6.2.1 Linear Response

In the simplest conceptual model of a response function, a genotype would have a linear response to an environmental variable. While such a response is physiologically unlikely to exist, a linear approximation over a small environmental domain may be satisfactory. Then, other genotypes could be expected to differ, at least in the level of the response line if not in the slope. In such a case, it would then also be true that, for any two environments, the genotypic ordering would be the same and the differences between them would be constant. There is no $G \times E$ interaction, and the model of independent genetic and environmental effects, $P = G + E$, would be estimable, even if it really is not meaningful to speak of a genotypic effect in the absence of an environment and vice versa.

For some traits, this is a fair approximation within the environments sampled. For juvenile height growth in Douglas-fir (Yeh and Heamon, in press) for the first 7 years, the ranking and size differences among families is reasonably constant over 11 planting sites. However, by the eighth year, interactions begin to appear. In the linear analysis of variance for such experiments, a general source of environmental or planting site variance can be identified and, in addition to the family effects, a general environmental × genotype "interaction" can be defined and estimated. It is also possible to partition the site variation into any identifiable environmental factors such as elevation or soil factors, and to similarly partition the $G \times E$ into elevation × genotype interactions, etc. Furthermore, if the genotypic source of variance is decomposable into family types or additive and nonadditive gene effects, then the G × E can be partitioned into additive × elevation, etc. (Namkoong, 1979). Such analyses of linear models can thus generate tests for departures from uniformity of genetic or environmental effects.

The concept of stability analysis as developed by Finlay and Wilkinson (1963) and Eberhart and Russell (1966) is based on the foregoing linear models, but focuses on differences in the slopes of the response functions. It also is based on measures of an environmental scale that is determined by the mean of the sampled genotypes rather than by any independent direct measure of an environmental variable. In an ANOVA with E environments and G genotypes planted on each, E − 1 independent environmental means and G − 1 independent genotypic means can be estimated. It is then also possible to estimate a linear regression slope for each genotype on the environmental means and to test if the coefficients are the same. If they are different, then regressions greater than + 1 are considered to be highly reactive genotypes that respond with a relatively great depression to poor sites and a relatively great superiority to good sites. Genotypes with slopes < + 1 are considered to be relatively stable. The selection procedure would then presumably involve some form of weighted mean yield as one trait and stability as another, in contrast to the approach taken in Section 6.3.

For some plant species, however, the highest average yield is associated with instability (Tai, 1971), and different environments may induce different kinds of genetic variance to be displayed by the same organisms (Perkins and Jinks, 1968). Then, uniformity of performance is most readily produced by different genotypic mixtures.

It would clearly be advantageous to use genotypes that uniformly perform best in all sites. There is some indication that such phenomena may exist, but to the extent that such uniform goodness does not exist, some compromise is required between maximum adaptedness with special breeds and limits to the number of special populations that can be specified for geographic areas or other site restrictions. Thus, preliminary surveys of the dimensions and extent of environmental variations are desirable to test the form and importance of these genotype × environment interactions as well as any changes in the genetic variances (King, 1965; Ledig, 1970; Squillace, 1970). Descriptions and classifications of genotypes according to similarity of responses can be useful to determine the existence of subsets of environmental variables and subsets of genotypes (Hanson, 1970). Regression types of analyses of genotypes on environments can greatly benefit both the analysis of forms of interactions and their practical use in breeding (Perkins and Jinks, 1968; Freeman and Perkins, 1971; Pooni and Jinks, 1980).

Differences in degrees of stability and response to favorable and unfavorable environments have been found for slash pine (Snyder and Allen, 1971). The only major difference between these methods and standard regression analyses is that the environments are scaled according to the average performance of the genotypes and not by known, measurable variables. Freeman and Perkins (1971) constructed an ANOVA for t genotypes (with $t − 1$ degrees of freedom), s environments (with $s − 1$ df), and an interaction (with $(t − 1)(s − 1)$df). The $s − 1$ df for environments are partitioned into 1 df for a sum of squares from an average linear regression and $s - 2$ df for the remainder. The interaction is partitioned into sums of squares resulting from genotypic differences in their linear regressions (with $t - 1$ df) and the remainder (with $(t - 1)(s - 2)$ df). Eberhart and Russell's (1966) partitioning is slightly different: The $s - 2$ df of the remainder from environmental regression and the $(t - 1)(s - 2)$ df of interaction remainder are used to construct a sum of squares of deviations of environments from linearity with $s - 2$ df for each genotype. Both methods, however, are essentially similar in seeking linear regressions and ANOVAs for testing those models.

For more general analyses of nonlinear and multiparameter responses, the same breakdown of degrees of freedom for multiple regression can be followed. For such cases, unbalanced designs with several environmental measures may find greater use, especially for forest trees with complicated response patterns. These experiments would require a careful allocation of possible treatments or environmental degrees of freedom into specified treatment combinations.

Although special treatment combinations can be most efficiently designed for purposes of regression analysis as given, or in more complicated response surface estimations, it will seldom be possible to test all genotypes on all site and silvicultural variable levels. The location of replicated treatment combinations around maximum response and maximum curvature zones, as previously suggested, may not be feasible. More often, it will be necessary to specify standard environments that can test the range of responses of interest or to specify indicator genotypes that provide some idea of the existence and form of genetic interactions for given site variations. In addition to the effects on selection, the existence of interactions causes bias in estimating genetic variances of single-location experiments, as previously discussed. However, even if the component of variance from interactions is small, its effect on selection can be significant (King, 1965). As a minimum program for testing, at least an average site would have to be sampled by all genotypes. More generally, to the extent that seedlings are available for testing, environmental factor combinations representing the breeder's best guesses on major site subdivisions should be sampled. If major genotype × site interactions exist, then selections can be made for specific sites for special breeds or for a single, generally useful breed.

Site sampling should follow the general principles of Box and Lucas (1959) to span as many site dimensions of present and future utility as possible. If balanced designs incorporating all genotypes across all selected sites cannot be installed, then unbalanced designs will find some utility. Recognizing that some interactions of special genotype–site combinations will not be observed, and if mean estimation is still of major importance, then partially balanced factorials of genotypes × sites can be profitably used. These may either involve a single, completely connected, partial factorial in which some overlap between genotypes and site combinations exists so that complete least squares of main effects at least can be determined or that some sub-blocking can be used instead. If genotypes are sub-blocked to gain experimental (mean estimation) efficiency, some loss occurs in that genotypes in different subblocks cannot be directly compared. The general design and analysis problems considered in Chapter 5 are directly applicable to these problems.

6.2.2 Nonlinear Response

Alternatively, Knight (1970), Baker (1987), Roberds and Namkoong (1986), and Gregorius and Namkoong (1986) believe that much information is lost when responses are actually nonlinear. The forcing of a linear fit requires estimating parameters that do not reflect relative environmental sensitivity and do not provide prescriptive information for selecting genotypes to fit environmental distributions.

A major general concept of interaction was proposed by Gregorius and Namkoong (1986, 1987) in which the general form of the response functions may be nonlinear and the genotypic differences need not be only constant or a multiplicative function of the environment as required by the analyses given. While generally applicable to continuous and discrete environmental classes, the implications for monotonic continuous response functions are comparable to the uses made of the stability analyses. In these cases, interaction is found when the response functions intersect and rank changes occur. Otherwise, the theory uses the relationship among the response functions to define a joint operation of genotype and environment to produce the phenotype. Genotypic effects and environmental effects can be separated and estimated, and the operation that joins their effects can be described. We are basically concerned with a set of phenotypic observations on a set G of genotypes in a set E of environments, and we can then hypothesize the existence of a phenotypic response function, ϕ, which specifies, for each pair (g, e) of genotype g (from G) and environment e (from E), the phenotypic expression $\phi(g, e)$. Hence, we seek the mapping for ϕ that takes values in the set F such that

$$\phi: G \times E \to F$$

describing the response of each genotype in each environment.

It can be noted from any given genotype g_i that its response to an environmental set, or its reaction norm, is an expression of the environmental effects for that genotype. If these effects were consistent, that is, if they would not change with the genotype, then as a minimum requirement two environments e_1 and e_2 having identical effects on g_1 [i.e., $\phi(g_i, e_1) = \phi(g_1, e_2)$] should also have identical effects on the other genotypes [$\phi(g_i, e_1) = \phi(g_i, e_2)$]. Analogously, e_1 and e_2 having different effects on g_1 [i.e., $\phi(g_i, e_1) \neq \phi(g_i, e_2)$] should also be different in their effects on the other genotypes [i.e., $\phi(g_i, e_1) \neq \phi(g_i, e_2)$]. In this case it can be consistently said that an environmental effect exists that is separable from genotypic effects. However, if the environmental effects are identical for one but different for another genotype, then the environmental effects are fundamentally inconsistent and are not separable from the genotypic effects. We term this kind of inconsistency an interaction, and define *separability of environmental effects from genotypic* as: the response functions for any two environments from E are either consistently identical or consistently different, that is,

> For any two elements e_1, e_2 from E,
> either $\phi(g, e_1) = \phi(g, e_2)$ for all g from G   (6.1)
> or $\phi(g, e_1) \neq \phi(g, e_2)$ for all g from G.

Conversely, we can analogously define *separability of genotypic effects from environmental* if and only if the response functions (reaction norms) for any two genotypes from G are either consistently identical or consistently different, i.e.,

$$\text{For any two genotypes } g_1, g_2 \text{ in } G,$$
$$\text{either } \phi(g_1, e) = \phi(g_2, e) \text{ for all } e \text{ in } E \qquad (6.2)$$
$$\text{or } \phi(g_1, e) \neq \phi(g_2, e) \text{ for all } e \text{ in } E.$$

If separability is realized in both directions, we may speak of *mutual separability*, or simply separability of genotypic and environmental effects. *Interaction* is then defined as the failure of separability that occurs if either Equation 6.1 or Equation 6.2 fails. Thus we may encounter two types of interaction: (a) if the environmental effects are not separable from the genotypic, then at least one of the environmental effects interacts with the genotypic effects, and (b) if the genotypic effects are not separable from the environmental, then at least one of the genotypic effects interacts with the environmental effects.

Obviously, there is a broader class of genotypic response functions that can be included in a set and still be called noninteracting, including nonquantitative and discontinuous functions. They are intuitively reasonable to define as being joint products of the two factors that are separable and hence "noninteractive." Clearly, the additive mode of action is a special case of a set of straight-line, parallel functions. Many other joint effect functions such as multiplicative or exponential joint functions are definable as particular modes of action between separable environmental and genetic effects. In the additive case, the effects can be transformed into functions that can be combined in a linear form. We now seek to define these transformations of separable functions in order to discriminate among forms of joint actions.

When a separate ϕ exists in the sense of Equations 6.1 and 6.2, it is desirable to then derive functions of genotypes which are in some sense "independent" of environments and vice versa. Thus for any ϕ that satisfies Equations 6.1 and 6.2, we would like to define genetic (δ) and environmental (ε) contributions that are not functions of the other variable and which can be combined in some form of mathematical operation (Ω) to form ϕ. That is, we wish to decompose ϕ into an operation (Ω) on mutually separable genotypic and environmental contributions (specified effects) and to be able to clearly define $\delta(g)$, $\varepsilon(e)$, and $\Omega(\delta, \varepsilon)$. For these purposes consider two functions

$$\delta: G \rightarrow C_\delta \quad \text{and} \quad \varepsilon: E \rightarrow C_\varepsilon,$$

where the contribution of each genotype as an element from C_δ takes the form $\delta(g)$ and is combined by the operation Ω with a contribution from C_ε for each environment to completely map the joint response function. Thus

$$\Omega: C_\delta \times C_\varepsilon \rightarrow F$$

and it obeys the relation $\Omega[\delta(g), \varepsilon(e)] = \phi(g, e)$. This then corresponds to the characterization of the response by ϕ for mutually separable effects. Neither effect should obscure the effect of the other and, therefore, for mutually independent contributions, we require that

$$\Omega(x, y_1) = \Omega(x, y_2) \quad \text{only if } y_1 = y_2 \qquad (6.3)$$

$$\Omega(x_1, y) = \Omega(x_2, y) \quad \text{only if } x_1 = x_2 \qquad (6.4)$$

For each x from C_δ and y from C_ε, $\Omega(x, .)$ and $\Omega(. , y)$ can be considered as functions.

$$\Omega(x, .): C_\varepsilon \to F \quad \text{and} \quad \Omega(. , y): C_\delta \to F$$

Then the conditions of Equations 6.3 and 6.4 are equivalent to $\Omega(x, .)$ and $\Omega(. , y)$ being one-to-one functions and thus invertible. Therefore, we only need to know what the operation is and either a $\delta(g)$ or an $\varepsilon(e)$ to derive the other.

Thus, if ϕ is separable, and therefore no interaction exists, there is always some operator on a genetic contribution function and an environmental contribution function that can be derived. Further, each genotype and each environment in the set can be uniquely characterized by its contribution function, which can be derived. Also, each genotype and each environment in the set can be uniquely characterized by its contribution function taken with respect to an appropriate average or other reference genotypic or environmental sample. We can therefore decompose the ϕ representation into genetic and environmental component contributions [$\delta(g)$ and $\varepsilon(e)$] and can describe how the separate functions jointly operate (Ω) in what we are now calling "noninteractive" conditions. Not only are these intuitively pleasing concepts of independent effects within a system of joint operations, the definitions are operable.

Thus, the Ω representation of response functions lays bare the essential way that a genetic contribution and an environmental contribution jointly operate to produce the phenotypic responses. The separable contributions themselves may be some complicated functions of the genetic, g, or environmental, e, effects.

As an example, we will consider a data set on seed source trials of *Pinus caribaea*, (Gibson, 1982) that are derived from replicated trials in several locations. We extract the data on stem straightness scores for seed source means in five of the test planting sites (Table 6.1). By analyzing the sources Alamicamba, Santa Clara, Guanaja, and Potosi, the function which characterized the joint property of source and environment seems to reflect an exponential type of operation (Fig. 6.1a). This suggests that the straightness scores for these source populations are jointly exponential effects of the source genotypic contributions and the planting environmental contributions. However, as can be seen in Fig. 6.1b, the two other source populations tested, Brus Lagoon and Poptun, do not have the same kind of joint operation and a linearity of genotypic and environmental effects with respect to each other is apparent. The interaction between these two sets of source populations lies not only in the difference in level of their reaction norms, but in patterns by

Table 6.1. Stem straightness scores for six sources in five environments

Planting environment	Seed source					
	ALA	STA	GUA	POT	BRU	POP
Chati	3.8	3.7	3.0	1.9	4.2	3.3
Mariti	4.2	3.8	3.4	2.0	3.3	2.9
San Pedro	10.3	8.9	7.7	5.0	7.1	6.8
Cardwell	9.3	8.7	7.0	4.9	9.0	8.5
Bukit Tapah	10.5	9.4	8.8	7.7	9.1	8.0

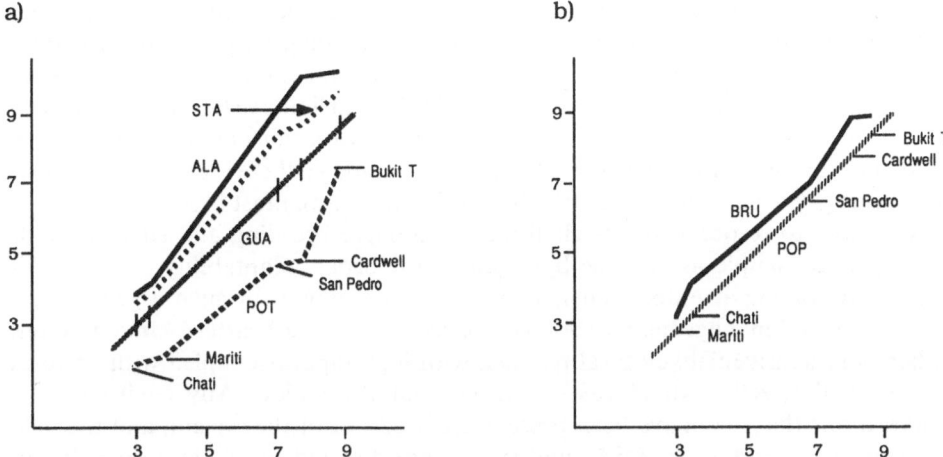

Fig. 6.1. Characteristic function for different sources and environments. (**a**) Four sources and five environments, which are exponential in form. (**b**) Two sources and five environments, which are linear in form. (From Gregorius and Namkoong, 1986.)

which one reaction norm relates to the others. Hence, there are differences in how environmental effects shape genetic responses.

In conclusion, if the concept and utility of the analysis of separable response functions is accepted and is clearly distinguishable from what we term interaction, we can apply the above analyses only within "noninteractive" sets of genotypes and environments. When our basic concept of consistent effects, as defined by Equations 6.1 and 6.2, breaks down, then we cannot define generally valid gene or environmental effects. However, if there are subsets of genotypes and environments within which consistency exists, then we can describe the amount of such subsetting that may be required and discuss other features of those subsets. It is also possible to explore the nature of relationships between subsets in order to jointly interpret some functions as possibly separable in a piecemeal fashion. It is not our intention here to further explicate the measures and interpretations of these interactions, but rather to extract as much interpretability as possible out of some basic concepts of noninteraction.

Although statistical procedures for estimation of these effects have not yet been developed, analyses of genetical-environmental effects have been made (Glock and Gregorius, 1986). In Section 6.3 we consider how the specification of genotypic response functions can be used in a selection program for the whole response function.

6.3 Changes in Environments and Genotypes

Breeders are faced with uncertainties in the future environmental distribution of forests, silvicultural regimens, and the kinds of genotypic response functions appropriate to breed for those conditions. Given the existence of genetic variation in response to environmental variation, there is an effective economic and managerial lower limit to the fineness with which we can breed for adaptation. Because there are also finite upper limits to the breadth of environmental variations to which genotypes can adapt, there are also economic and biological limits to how broadly we can breed for adaptability. Further, because temporal variations exist in the physical and biotic nature of the environments we wish to manage, there is uncertainty about the appropriate environmental distribution for which we must breed.

One way to respond to this challenge is to make breeding and silviculture as managerially simple as possible by breeding for broad adaptability and by silviculturally forcing a uniform environment. The paradigm for such a genetic and environmental management system is tobacco, in the southeastern United States, where an intensive field cultural regimen with high input is coupled with uniform testing and selection standards set throughout the region. Any environmental interactions that may have been present are bred out of the system, and the field management systems are fairly uniform or are directed to create uniformly appropriate growing conditions; marketing is also essentially made uniform throughout the region. Thus, as many economic and environmental sources of variability as possible are eliminated, and genotypic performance is made uniform.

When markets are more weakly controlled, multiple products exist, and silvicultural systems exercise far less intense control of both physical and biotic environmental variables, there is a question as to the possibility that such broad, regional, uniformities are either possible or desirable. Markets do change; insects and pathogens are not intensively managed; the forest environment is usually much more macroscopically variable; and management is often directed for diverse objectives in any case. Thus, for few forest trees would it seem feasible to even consider the tobacco paradigm. The primary question then becomes one of how much genetic homeostasis is needed, and how should it be constructed. We consider the question of environmental subdivision or zonation in Section 6.3.1 and how we might structure the genetic variation in Section 6.3.2.

6.3.1 Ranking Change, Response Function, and Zonation

Although such populations are unlikely to occur in forestry, single-population breeding can be the most economically feasible breeding option for species with well defined and constant regimens of adaptability over which the genotypes do not change rank to any important degree. If the species is also of relatively low economic value and is likely to remain so for any reasonably expected changes in the economic or ecological environment, then the only possible reason for carrying more than one population would be for replicate population breeding (Baker and Curnow, 1959).

However, a more critical point to consider in most breeding situations is that, regardless of the current priority in breeding, different present or future uses and different environmental adaptability requirements will necessitate a diversity of genotypes and breeding for an array of economic and/or ecological abaptability. Therefore, with limited adaptability of available genotypes in any one trait of interest, population subdivision may become the most economically feasible breeding approach. Obviously, not all traits have to display a genotypic change of rank to consider developing such subdivisions, one will be sufficient if it is economically important. In that case, the other traits may be selected for in the same way in the subpopulations that are segregated for environmental adaptability. Alternatively, one may breed for the same general adaptability and then segregate subpopulations for a specific trait–environment subdivision.

Even if there is no change in genotypic ranking in any component individual trait, a change in ranking can occur in a multitrait selection index. That is, if an analysis is made of several traits and none shows any genotypic change of rank over the environments tested, there still may exist a multitrait index in which genotypic values do change rank over these same environments (Namkoong, 1986). It is obvious that such change in ranking of multitrait selection index occurs if the relative economic importance of two traits changes with changes in the environment (Namkoong and Johnson, 1987). Perhaps more surprising is the case in which the relative economic importance of the two traits is constant and the only deviation from the linear model is that one of the traits (A) displays a multiplicative effect of genotypic and environmental effects. That is, the environmental deviation from the mean environment is multiplied by the genotypic deviation from the mean genotype to produce the phenotype. While no change in genotypic ranking is observed within any environments, it is possible, by economically weighting the A trait heavily, that in an index of traits A and B there will be created an index value × environmental interaction (Namkoong, 1985b).

In general, tree response to different environments is likely to be more complex than these examples. The genotypic response functions are not just multiplicative or additive, and relative trait values change in more complicated functional relationships with the environment. Therefore, finding means of determining multiple optimal combinations of genotypes and environments is an important process in tree breeding.

If such complex changes in genotypic ranking exist in individual traits, and if the responses of the individual genotype to an array of environments can be approximated by a bell-shaped Gaussian curve, then it is possible to define not only an optimal zonation for subdividing the environment, but also to define the kind of genotypic response function that would maximize the value of multiple breeding populations (Roberds and Namkoong, 1986; Namkoong, 1980a).

We thus view the breeding problem as one of not only maximizing yield, but one of defining the environments and response shapes to use to raise the yield in the most economically efficient ways. Thus, the fact that genotypes and heritability may change over environments does not prevent us from developing a maximization theory and methods for constructing breeding populations. Our first breeding efforts may have needed to use arbitrary zonations, and selection was directed for an average environment, but this is neither optimal nor necessary.

Zonations were sometimes initially based on a "local-is-best" paradigm (Namkoong, 1969), with seed zones set up in terms of distance over which seed could be moved from its parental origin. This does not define an optimal zonation but instead restricts the development of zones to a presumed, not an experimentally estimated, response function, and does not use information on the frequency distribution of environments. Thus, it is neither biologically nor economically, based. Methods that use these data would be more amenable to gene management programs that maximize yield by using populations for targeted environments.

6.3.2 Zones and Multiple-Population Breeding

With more than one environmental (economic or ecological) zone it is obvious that different populations would have to be bred for different adaptabilities. For developing populations for future environmental distributions, wider adaptabilities among populations may be optimal to develop. It is also obvious, as discussed in Chapter 5, that multiple populations with different trait-selection goals could be a useful developmental strategy to follow. The genetic management problem then is how to accommodate multiple objectives with a finite number of populations. It might be conceivable to create a full complement of multiple-trait selection goals for each environmental zone. An identical set of populations would be developed if, for each multitrait cluster, environmentally zoned subpopulations were constructed. However, because the number of separate breeding populations multiplies rapidly with any increase in either the number of trait clusters or environmental zones, less complete sets are necessary to design for any applied (practical) multiple trait-selection program. Several ways to organize such multiple populations suggest themselves. For selected environmental zones, multitrait clusters can be targeted in subpopulations; and vice versa, environmental zones can be targeted for important trait clusters.

Alternatively, a single primary environmental zone and trait cluster may be identifiable in some cases, and then supplemented. In cases of high certainty that such a single population will be a good basis for future development, any multiplicity of populations would be tightly clustered around the same selection objectives and hence could be indistinguishable from the replicate populations of Baker and Curnow (1959). However, under high uncertainty at least three types of expansion design can be considered. One type would increase the number of trait cluster populations for an identified primary environmental zone, or increase an environmental zonation range for a primary trait cluster. A second type would simultaneously increase the span of both environmental zones and trait clusters so that the extremal corners of adaptation are developed. This might be most useful if breeding for both environmental and trait extremes is expected either to be directly useful, or intermediates in adaptability are expected to be usefully bred or hybridization of populations. Finally, if a primary population can be identified, and some uncertainty exists that the present primary population represents a reasonable average expectation of future needs, it would then be possible to construct wide or narrow circles or ellipses of optimally distal populations in which subpopulations are located.

In contrast to the possible subdivision of the overall breeding populations, we considered in Chapter 4 the possibilities of breeding in different populations for different kinds of gene action. In the present context of environmental variations it is conceivable that nonadditive gene actions may be more important in some environments than in others (Roberds et al., 1976). In such cases, it might be well worthwhile to use different breeding methods in different environmental zones. For example, if heterosis breeding is successful in extreme or stressful environments, but not elsewhere, then an environmental zonation and the choice of breeding method would be confounded. The simultaneous designation of zones of breeding methods would be complicated but at least could conceivably be optimized. More complex yet would be the case in which trait combinations, breeding method, and environmental zonation are all to be maximized simultaneously.

A similar approach to this three-dimensional problem can be taken as with the preceding two dimensions of traits and environments. Although such complexities might seem daunting to tree breeders, the fact that intrapopulation breeding can be well and efficiently done in relatively small SRS populations allows the breeder to consider the strategic development of the whole meta-population. Thus, it is still possible to consider a few tens of populations as providing the flexibility of breeding among populations, while maintaining effective population sizes of a few tens of individuals within each.

A substantial need arises for efficient testing and breeding on consideration of such environmental–genetic management. Especially when considering the need for estimating responses in more and more extreme environments than presently needed does the use of efficient field designs become increasing important. However, even given the capacity for great efficiency, the greatest need that we as geneticists have in this respect is for greater depth and sophistication in our understanding of physiology and silviculture.

6.4 Competition

One of the traditional ways in which foresters control an environmental factor is by spacing or controlling stand density at planting and by thinning or selective cutting during stand development, which affects intraspecific competition. Interspecific competition in the form of herbaceous or shrub competition during early growth stages may also be considered as a form of competition. These forms of density-dependent competition may be thought of as mechanistic effects of shading, which limits incident light, or root competition, which limits access to soil resources, and hence are merely biotic instruments that affect physical environmental variables. In some cases, the physical effects may be simple, direct effects, easily mimicked by such means as mechanical shading or soil moisture control or by controlling soil fertilizer levels. If such simple, testable effects are well correlated with stand density responses, then direct means for testing and predicting performance over a wide range of environmental conditions are available. It would then be possible to select for density response functions that would maximize yield for any desired range of spacings. Such experiments have been conducted for several agronomic species, but not thus far for forest trees.

Physical effects may not be so simple, however, and may involve, for example, the time at which moisture stress occurs, or the sequences and combinations of events that occur in fields of competitors. In such cases, although density dependence is still operating, there may be no recourse except direct field testing under different spacings. Two kinds of density response designs have been suggested for forest trees; rectangular plots with a range of densities in different spacing blocks (espacement), and row plots in "Nelder's" designs with spacing changing within plots. In the rectangular plots, the blocks must be of different physical dimensions (and therefore include different amounts of microsite environment variation) or include different numbers of trees per plot. Furthermore, they are difficult to design if there is any desire to test for differences in competition among specific genotypic pairs.

The Nelder's type of design is a row plot with plots arranged as spokes on a wheel and environmental blocks appearing as large wheels. Spacing is closer at the center and wider at the outer circumference, and for the same total area, or for the same number of trees, more pairwise combinations of genotypes can be tested (Namkoong, 1965). Each plot provides an estimate of the density response function, but the estimate must be made with heterogeneous errors because the trees at wider spacing are affected by more widely varying soil differences as well.

In both cases, the average response function could be an estimate of a general response to all the physical events that affect tree growth as constrained by other competitors for the same physical resources. In either case a general density response function could be selected for and expected to be consistent regardless of the specific competitors it met.

Beyond such physical effects, there may also exist important effects that may not be reducible to particular environmental factors or to any general, consistent, average response function. Instead, competition may be idiosyncratic in that the outcome may be dependent on the particular competing genotypes, which cannot be related to competition for light or water, or may not even be consistent in competitive ordering. That is, A may outcompete B, which can outcompete C, but C may still outcompete A. Although such effects may result directly from physical factors, the complexity of biotic interactions may also introduce frequency-dependent effects. In that case, the rank ordering of genotypes can change depending on the relative frequency of competing genotypes, and hence the spacing effect can be overridden by changing the frequency of competitor genotypes.

In the case of simple density dependence, whether reducible to single physical factors or not, the competing genotypes were assumed to have net effect on the ranking of genotypes over a density interval. In more complex biotic interactions that may also involve interactions of physical factors, genotypic rankings may change with the frequency of competitor types, even at the same densities. The effect is what Sakai (1955, 1965) defined as intergenotypic competition when he observed that at a standard density, the performance of rice plots varied as the percent of its competitors switched from 100% of its own variety to 100% of another variety. This was also observed for soybeans by Schutz and Usanis (1969) and Schutz and Brim (1971), and for forest trees by Adams et al. (1973), Sakai and Mukaide (1967), and Sakai et al. (1968).

Under these models, a breeding question is then raised with respect to the existence of general competitive abilities and optimal ways to select and breed for maximum yields (Allard and Adams, 1969) and the existence of evolutionarily stable genotypic frequencies in forests (Hühn, 1970). In terms of how selection affects the composition of competitive interactions, if the genes which affect competitive ability are quantitatively inherited, Griffing (1967) has developed a model that is parallel to his classical model of selection effects. The allelic combinations that result from truncation selection and from reconstituting a breeding population by crosses among selected individuals, are traced. In definitions similar to Hühn's (1969), Griffing assumes that alleles have direct effects on their own genotype's growth as well as associate effects on those with which the genotype competes. Thus, instead of a simple $d_{ij} = \alpha_i + \alpha_j + \delta_{ij}$ genetic model, he defines $_{i,j_1}d_{i_2j_2}$ as being the genetic value of individual 1 in the presence of individual 2, and hence for populations of size 2, the array of genotypes with allelic frequencies p_i and p_j is

$$\sum_{i,j} p_i p_j (A_i A_j) \times \sum_{i,j} p_i p_j (A_i A_j) = \sum p_{i_1} p_{j_1} p_{i_2} p_{j_2} (A_{i_1} A_{j_1} A_{i_2} A_{j_2})$$

Genotype value $A_{i_1} A_{j_1}$ as expressed in $(A_{i_2} A_{j_2}, A_{i_2} A_{j_2})$ is $_{i_1 j_1}d_{i_2 j_2}$,

$$_{i_1 j_1}d_{i_2 j_2} = {}_d\alpha_{i_1} + {}_d\alpha_{j_1} + {}_d\delta_{i_1 j_1} + {}_a\alpha_{i_2} + {}_a\alpha_{j_2} + {}_a\delta_{i_2 j_2} + {}_{da}(\alpha\alpha)_{i_1 i_2}$$

$$+ {}_{da}(\alpha\alpha)_{i_2 j_2} + {}_{da}(\alpha\alpha)_{j_1 i_2} + {}_{da}(\alpha\alpha)_{j_1 j_2} + {}_{da}(\alpha\delta)_{i_1 i_2 j_2}$$

$$+ {}_{da}(\alpha\delta)_{j_1 i_2 j_2} + {}_{da}(\delta\alpha)_{i_1 j_1 i_2} + {}_{da}(\delta\alpha)_{i_1 j_1 j_2} + {}_{da}(\delta\delta)_{i_1 j_1 i_2 j_2},$$

where

$_d\alpha_{i_1}$ = direct additive effect of allele A_{i_1},

$_d\delta_{i_1 j_1}$ = direct dominance effect of $A_{i_1} A_{j_1}$,

$_a\alpha_{i_2}$ = associate additive effect of A_{i_2} as measured on $A_{i_1} A_{j_1}$,

$_a\delta_{i_2 j_2}$ = associate dominance effect of $A_{i_2} A_{j_2}$ as measured on $A_{i_1} A_{j_1}$,

$_{da}(\alpha\alpha)_{i_1 i_2}$ = additive × additive interaction effect between direct allele A_{i_1} and associate allele A_{i_2},

$_{da}(\alpha\delta)_{i_1 i_2 j_2}$ = additive × dominance interaction between direct allele A_{i_1} and associate genotype $A_{i_2} A_{j_2}$,

$_{da}(\delta\alpha)_{i_1 j_1 j_2}$ = dominance × additive interaction effect between direct genotype $A_{i_1} A_{j_1}$ and associate allele $A_{i_{t_2}}$, and

$_{da}(\delta\delta)_{i_1 j_1 i_2 j_2}$ = dominance × dominance interaction effect between direct genotype $A_{i_1} A_{j_1}$ and associate genotype $A_{i_2} A_{j_2}$.

These interaction effects are not epistatic effects but average intergenotypic effects caused by allelic effects in the sense of affecting competitive phenotypes. The total genotypic variance of $_{i_1 j_1}d_{i_2 j_2}$ is

$$\sigma_G^2 = {}_d\sigma_A^2 + {}_d\sigma_D^2 + {}_a\sigma_A^2 + {}_d\sigma_D^2 + {}_{da}\sigma_{AA}^2 + {}_{da}\sigma_{AD}^2 + {}_{da}\sigma_{DA}^2 + {}_{da}\sigma_{DD}^2$$

and the covariance between direct and associate effects is

$$_{da}\sigma_A = 2\sum p_{i_1} ({}_d a_{i_1}) ({}_a a_{i_1})$$

The consequences of ignoring the existence of interactions among competing genotypes in selecting individual trees can then be traced with the same simplifying assumptions that Griffing used to derive the noninteractive solution. However, by also tracing the associate effects of the trees selected, he derives a selective value of

$$W_{i_1 j_1} = 1 + \left(\frac{s}{\sigma^2}\right)_{i_1 j_1} d$$

and the gametic array of selected individuals is

$$(\tfrac{1}{2}) \sum p_{i_1} p_{j_1} w_{i_1 j_1} (A_{i_1} + A_{j_1})$$

Random mating among these selected parents then generates a new mean of approximately

$$\sum p_{i_1}p_{j_1}p_{i_2}p_{j_2}\left[1 + \frac{s}{\sigma^2}\left({}_da_{i_1} + {}_da_{j_1} + {}_da_{i_2} + {}_da_{j_2}\right)\right]({}_{i_1j_1}d_{i_2j_2})$$

$$= \frac{s}{\sigma^2}\left[{}_d\sigma_A^2 + {}_{da}\sigma_{AA}\right]$$

This reduces to the familiar sh^2 if the covariance of direct and associate effects is zero, which implies that there is independence of the two effects and that, on the average, we can select on individual performance with impunity. However, if a strong competitor is a vigorous tree that suppresses its neighbor, then a negative covariance can exist and gain can be substantially reduced. If the covariance is positive, a benefit is obtained over the case of ignoring competition effects.

If some form of group selection is used in which groups of size (gr) 2, 3 ... n are chosen for mixed growth properties, the selective value of groups is taken as the contribution of both direct and associate effects. Then under group selection

$$w_{i_1j_1,i_2j_2} = 1 + \frac{s}{\sigma^2}(gr)\frac{1}{2}\left({}_{i_1j_1}d_{i_2j_2},{}_{i_2j_2}d_{i_1j_1}\right)$$

and selection of groups within which random mating occurs yields a mean of approximately

$$\frac{1}{2}\frac{s}{\sigma^2}(gr)\left[{}_d\sigma_A^2 + 2_{(da)}\sigma_{AA} + {}_a\sigma_A^2\right]$$

While the latter factor can never be negative and never smaller than σ_A^2 in the individual case, the second factor can be small depending on testing ability.

Extending these results to groups of size n can benefit selection. However, if testing is limited, it might be easier to use direct individual selection with separate measurements such as crown diameter, root exudates, etc., to establish the form of competitive influence and to compose populations without direct group testing. This is essentially the recommendation of Toda (1956), who recommended selection for growth with narrow crowns in *Cryptomeria*.

Obviously, for forest tree breeding, when changes in both density and genotypic composition of competitive abilities are occurring we must breed with little knowledge of how the density response curves might change with specific competitors. It seems likely that these effects will be age dependent and hence even more complex than presently understood. However, if pure density response is independent of or positively correlated with intergenotypic competitive ability of a complementary type, then the two kinds of responses can be simply treated as any two traits for selection purposes. With further research on intergenotypic effects and how density may affect those relationships, we are liable to learn much more about the complexities of growth among individuals at the population level of biotic effects. It might also be possible to select for types of density and intertree effect such that our yields are improved more than is possible by just considering these effects separately. That is, the interaction of those two effects may prove more useful to develop than either factor alone. "Competitive ability" is obviously not a simple, single trait.

Thus far, tree breeding programs cannot use competition as a selection factor to increase our yields (Nance and Wells, 1981). Thus, in spite of the silviculturists, control over plantation composition and spacing, our inability to test and select for competitive ability remains. Instead, breeding programs are largely conducted for testing under some assumed standard spacing and general competitive mixture of genotypes. For these objectives, testing in single-tree or short-row plots among candidate genotypes or families is a reasonable design to adopt, even though it may yield little information for research on competition. Early-generation selection may have little effect on developing populations for other density conditions or for particular kinds of intergenotypic competition. For breeders to be able to use these kinds of effects and perhaps for silviculturists and breeders to design future forests for a mutual genotypic and ecological fit, initial population sizes must be kept large enough that future changes in tandem selection can be effective. The selection of genotypes for different conditions that can be silviculturally managed may be an important emphasis in future generations of breeding and would have to be considered at this time to be largely independent of present selection emphases. If negatively correlated, then larger population sizes are required to maintain the useful alleles for future selection. However, if independent then any general progress we initially make can be cumulative only if the selected populations of the second and third generations of breeding are large enough that essentially new selection efforts can be effective.

CHAPTER 7

Coping with Present and Future Breeding Problems

It is clear that breeding progress has been made in many species on many traits, and can continue to provide forest managers with new varieties that are capable of more different performances than presently exist. While there are exceptions and potential problems that may be genetic, economic, environmental, or managerial in nature, there are established techniques for avoiding or overcoming many such limitations. In the previous chapters we discussed long-term breeding techniques that are not yet adopted or are just being adopted by breeding organizations around the world. Although most commercial tree species have long generation turnover periods, we take it for granted that tree breeding activities will continue into the foreseeable future and that contemporary breeders can afford to plan and develop multiple-generation tree breeding systems. The main objective in this chapter is to discuss strategies for dealing with future uncertainties and multispecies breeding strategies. To develop such strategies, however, breeders need means of coping with problems inherent in tree breeding. This chapter is also intended as a bridge between the breeding problems discussed in the previous chapters and conservation-related issues and problems that are discussed in Chapter 8.

In Section 7.1 we discuss means of coping with difficulties associated with using various gene actions, multiple-trait objectives, and multiple environments. In Section 7.2 we discuss breeding when different levels of uncertainty exist and discuss means of developing breeding strategies under such uncertain and variable ecological and economic environments. In Section 7.3 we discuss multiple-species breeding strategies.

7.1 Coping with Inherent Breeding Problems

In Chapters 3 through 6 we discussed various breeding techniques. We discussed these in turn as a series of increasingly difficult problems. In this section, we assume a tree breeding organization, with limited resources, in which the breeders have identified and classified different potential opportunities and problems in breeding. For example, breeders are assumed to have determined the traits to be bred, and to know the types of gene action associated with traits to be bred. This assumption may not hold for some breeding organizations, especially those which are newly being developed. In this case, the discussion in this section can serve as a guide for developing research and breeding strategies.

7.1.1 Simple Recurrent Selection

In the simplest case, the genotypes for a breeding population are viewed as a simple collection of a large number of independently and largely additively acting alleles. They are intermated in some balanced or "random" pattern such that each parent has an equal gametic representation in the progeny generation. Selection is made at one stage in the life cycle, on the basis of a single ordering of genotypes made on one trait or a function of several trait values, for a single environment or a distribution of environments. While no single gene or set of genes may operate in such a way, with a large enough set of genes there would be a shift in the average frequency distribution, and hence heritable genetic gain is achieved. Also, for a large enough set of loci, in a large enough population with many such marginally effective alleles, low-frequency alleles may be maintained and eventually prove useful in advancing genotypic breeding values. In addition, with selection for specific performances, any mutations that might be useful would be among the lowest frequency alleles that would be saved and ultimately used (Hill and Rasbash, 1986a). Thus, even with inexact knowledge of single gene effects and which alleles at which loci may be useful in breeding, the effects of selection can be effectively approximated by quantitative genetic models.

However, given the uncertainty of the approximation of the model to reality, some caution is needed before reliance is placed on the effectiveness of selection in small populations (say a number of less than 20). For even such populations, however, the estimates of additive genetic variance may be reasonably stable, and iterated gains as approximated by heritability (Chapter 5) can be perpetuated in simple recurrent selection (SRS) procedures.

Even within a SRS context, however, it is possible to consider using a highly selected subset of the parents in any one generation to generate not only the commercially used seed, but also to constitute a mini-breeding population with less genetic variation. We can conceive of elite lines thus being developed and formed into a synthetic variety or used as single-cross parents in a hybrid breeding program. In that case, the SRS population becomes a less intensively selected, but steadily improving, base population from which lines can later be drawn as needed. At a slightly more remote level, we can conceive of the SRS population as deriving from a larger base population that is either held in some storage or only mildly selected for some general adaptability. These systems in fact are hierarchies of populations that agronomic crop breeding practices have used with some successes (Kannenberg, 1984). For species with high economic value and rapid generation breeding, it is possible to support such programs, but even for many of the annual food crops, such systems are difficult to maintain (Namkoong, 1984a). Nevertheless, even for relatively simple systems, there may be a need to consider these population structures.

7.1.2 Nonadditive Gene Action and Hybrid Breeding

Other than the SRS systems, gene actions may be such that nonadditive gene actions are important enough to use for some traits or for general vigor in all traits. As a breeding objective, cloning ability is difficult to develop, but as a means of using nonadditive effects and as a means of improving by testing, it offers substantial benefits. One way to use nonadditive effects in breeding would be to select for disequilibrium effects in a hybrid breeding system, or to select for families that show unusually high intrafamily segregations from either developmental instabilities or such genetic phenomena as high frequency of mobile elements.

Otherwise, several levels of hybridization can be constructed. It is often possible to select particular specific family crosses for commercial seed or for clonal propagation, even within a general SRS program. Any such particular specific combining ability or hybrid performance would be an extra benefit in performance, much as any special temporary boost in forest productivity might be that is built anew in each generation on top of the basic gain achieved in the SRS population. Only if the parents were to be developed into parental lines by selfing or some form of inbreeding could the specific combining abilities be accumulated. While worthy of examination as an alternative breeding system, for reasons stated in Chapter 4 this seems an unlikely way for forest tree breeding to develop. Rather, if hybrid effects are to be useful, they more likely will be found at the population level of crossing where some form of RRS may be practiced. Similarly, interspecies crosses may prove to be a useful basis for developing RRS systems, but here the evidence thus far for forest trees indicates that the value of hybridization lies in its combining of traits and not in overdominance types of gene action. Hence, in these cases the long-term genetic development of populations will often be most easily achieved by using hybrids as a basis for SRS systems or RS for General Combining Ability (GCA), rather than for heterosis breeding per se.

The question of the importance of heterozygous effects remains an open question in tree breeding. On the one hand, Namkoong and Bishir (1987) indicate that lethal alleles affecting germination are largely recessive mutants at low frequency, which can appear at many loci but can be easily purged. On the other hand, persistent increases in heterozygosity as tree populations age indicate that some heterotic effects may be important even if they exist at few loci (Bush et al., 1987). Thus, either specific crosses at the individual or population level should be made to maintain heterozygosity, or at least sufficiently large breeding populations should be maintained and intercrossed to maintain heterozygosity.

In all these cases, sufficiently large populations are necessary to permit recombination and the regeneration of genetic variations. All these programs can be conducted with the presently available technology and the genotypes can be organized into large single, or several replicate or paired hybrid parentage populations. While hybrid breeding requires progeny testing, all these programs can be successful with little testing effort and with a simple operational plan. For many important species, however, problems do exist or may soon appear in selection and mating and require other than the simplest breeding plan.

7.1.3 Coping with Low Heritability

Obviously, many traits may not show much progress in such systems, not because the gene actions and population size are incorrectly modeled but because simple recurrent selection is inefficient. Greater investments of time and experimental outlays into developing aids to selection may be an alternative means for enhancing the breeding system's value. If juvenile selection is poor then we can either delay selection and extend the breeding cycle, initiate tests in effficient designs, select on correlated traits, or simply accept a lesser rate of gain by following SRS without special testing. Of these options, efficient experimental designs and indirect (juvenile and correlated trait) selection designs would improve gain rates and provide additional information for improved selection procedures in subsequent generations. However, they would be costly of time and effort, and may have to be repeated in some populations at regular intervals. Therefore, while options clearly exist for achieving gains in spite of low heritabilities, they require some investment.

In the initial generations of tree breeding when there is little knowledge about heritabilities of any traits and experience in handling experimental materials is limited, selective breeding is necessarily conservative. However, if information is available on heritability or can reasonably be inferred from other populations, then it can be feasible to either forego testing or to shift selection objectives until better information is available. The latter option involves breeding on those traits that are amenable to SRS in this generation and developing tests such that correlated traits or other selection methods can be used in the next generation on traits with presently low heritability. Thus, for some traits, improvements in testing, predictability of performance, or better control of environmental variability can improve heritability and make selection on them more economically effective while delaying their inclusion in the selection criteria. In general there will always be a problem in evaluating traits for inclusion, if they are costly of either time or testing effort. Such costs should be included as a negative effect on the economic benefits of trait improvement. Because the costs of progeny testing are often very high and the marginal gain improvement relatively small, even for traits for which heritability can be substantially improved (Chapter 5; Namkoong, 1979), a reasonable conjecture is that for many traits SRS will be good enough. However, since the value of experimentation to future breeding is large and can change heritabilities and juvenile–mature correlations, the research objectives of progeny testing are expected to be worthwhile. When possible, therefore, selection should concentrate on traits of high value with inferred good heritabilities, and research should concentrate more on advantages in subsequent generations than on improving gains in the current generation.

7.1.4 Coping with Multiple Gene Actions

In Section 7.1.2, the problem of breeding with mixed additive and dominance gene actions among loci for the same or different traits was discussed. Another type of gene action mixture that may be encountered is that in which some traits are essentially controlled by one or a few loci of large effect, while others are more truly quantitatively inherited. The traditional breeding methods for fixing the appropriate alleles of a qualitative trait have been well developed elsewhere (Allard, 1960; Mayo, 1980) and require no review here. If the favorable allele frequency is high enough in the initial breeding population, then an index or ICL selection procedure would provide gain in both types of traits. Problems with these methods enter only if the favored allele frequency is low or the correlations are strong and negative for the quantitative trait. If strong negative correlations exist, then an ICL procedure, first selecting for the qualitatively inherited trait and then selecting for the quantitative, would ensure success with both kinds of traits. Otherwise, by index selection there is a finite probability of losing the favored alleles of the qualitatively inherited trait. This problem is more serious if the favored allele of the qualitative trait is at low frequency. The methods of ICL would also be appropriate when a single trait has both major effect genes and many quantitative effect modifiers.

If the favored allele is effectively nonexistent in the target population, but exists in an otherwise undesirable population or species, then the hybridization and backcrossing procedures of traditional plant breeding would be needed. If the techniques of gene transfer become available for such genes, and the genes are well enough known, then DNA transfer may become an economically feasible means of widely introducing single foreign genes into target populations.

7.1.5 Multitrait Selection

In addition to the problems of coping with low heritability and mixtures of major quantitative gene effects in single-population breeding, multitrait selection introduces other kinds of selection problems that may require different breeding sequences. In the simplest case, the methods of index selection described in Chapter 5 would be sufficient for creating a single scale of value out of many traits, and this composite trait, the index, could serve as a basis for single-population breeding. However, the definition of a single "best" direction for the breeding population is of dubious value. Given the statistical difficulties of estimating a best set of weights, the problems of defining simple linear value functions, especially when traits are nonlinearly related to value and economically interdependent and uncertain, become virtually impossible to treat. Under such conditions, a multiple-population strategy for all the traits can be reasonable to develop. Instead of a single population to breed, several could be bred for differently directed trait combinations in a multiple-index selection strategy (Namkoong, 1976) aimed at diversifying trait performances among populations. Presumably, the forest manager may then choose which combination of populations to use in any planting, and the breeder would then increase the supply of seed from these populations or their hybrids.

For practical breeding programs, however, it seems obvious that including many traits (more than four or five), even in a subdivided multiple-index selection strategy (MISS), would often be counterproductive for effectively diversifying the populations. It might therefore be more manageable to consider subdividing traits into two groups; those about which sufficient biological and economical information is available to effectively use index selection, and those which might as effectively be used in a tandem or ICL procedure.

For selection within a single breeding generation, it is often possible to cull a population in a series of early screenings to reduce the size of the population that might have to be analyzed or progeny-tested in longer or more expensive tests for some traits. Early culling on juvenile performance traits, such as on early survival or juvenile diseases, can reduce extensive test costs for mature growth or other low-heritability traits. Some traits may take so long or be so expensive to evaluate that it might be better to forego selection on them until research can uncover means of indirect selection on correlated traits. Such tandem selection would depend on effectively raising the heritability of some traits by better testing or observation.

Finally, it is always possible to reduce the complexity of genetic management by simply assigning traits for improvement to different populations and hybridizing them for producing a variety with the combined traits. Although this is never as theoretically efficient as single-population index selection, the theoretical loss may be small, but the gain in being able to effectively manage the breeding may offset such inefficiencies. This assumes an additivity of gene action for each trait and an additivity of value among traits. If either is false, then the ICL procedures may in fact be more efficient than an index. We can thus conceive of traits that are more suitable for RRS than for SRS procedures, and could then employ SRS on an index within each of two populations but select for those traits with overdominance in a hybrid test for population intercrossing. Such combined SRS within RRS systems for different traits are merely trait-confounded variants of the systems discussed in Section 7.1.2.

Thus, even for difficult problems of population breeding, it is possible to establish practicable breeding programs. Although they may not be as simple as the simplest one-population, composite-trait, single-gene action type of SRS, they are possible to organize. If gene actions and traits are simple then there is little need in any one generation to do more than simply ensure that some minimal N_e is maintained. With more problems in mixed gene actions, low h^2, and mixed traits, or the fear that gains will be severely limited because these problems exist, then the value of more intensive breeding must be weighed against its costs in time, effort, and managerial expenses. The technical problems are not as limiting as the economic problems of willingness to invest in research and development, and tree breeders should become more concerned with developing breeding options than with the technicalities of seed or clone production.

7.1.6 Multiple Environments

If we simplify consideration of traits to a single trait or a single composite scale, it may be possible to examine the G × E and determine the value of subdividing populations, defining zones or target populations of environments (Comstock, 1977), and breeding within these for maximum value (Chapter 6). With new efficient experimentation, this is indeed far more feasible now than ever before. However, it is also possible, as discussed in Chapter 6, that G × E can be created by the very fact of the multiplicity of traits. This results either from nonlinear genotypic responses in some traits of value that affect an environment-by-genotype value function "interaction," or from a value function that changes with environmental variables, or both. The interaction effect is strong when different traits react to environmental variations in dissimilar ways, and hence require different breeding zonations if treated independently. In such cases, a single value function (constant in its parameters) for all traits may exist, and a single zonation can be determined.

Alternatively, breeding populations may be segregated into trait sets within environmental zones. The finest such subdivision would be the smallest environmental zone that contains no trait interactions, and could be constructed by independently optimizing environmental zones for each trait and breeding different populations for each of the intersections of trait zones. Because this might require more breeding populations than can be afforded, means for aggregating zones must be derived. One possibility is to segregate a small set of environmental zones, and for each segregate breeding populations for a small set of different trait objectives. To the extent that some traits have little environmental interaction, the selection goals would be the same in all environments. To the extent that little uncertainty exists as to trait value, the environmental zonation would be the same for all traits. To the extent that some traits would have little importance in some environments, the number of traits selected in each environment may differ. Several possible arrangements of such factorial combinations of traits and environments suggest themselves, including a boxing-in of the extremes and mean or median environments, with each zone repeated over multiple-trait objective sets. Alternatively, strategies for developing a "main" population for an average environment and an average trait function could include supplemental populations for more extreme environmental and/or trait combinations, in unbalanced factorial combinations.

Clearly, it is also possible to consider that breeding may well be directed to different environmental zones with different gene actions used in each. That is, for environmental zones in which dominance and overdominance types of gene action are important, a breeder may wish to use a hybrid breeding population. This may occur, for example, on more extreme sites such as found by Roberds et al. (1976), where the dominance genetic variance was larger than the additive genetic variance for loblolly pine on both nitrogen-poor soils and on overfertilized soils. If this is caused by overdominance gene effects, it would be feasible to consider hybrid breeding with some form of RS for SCA on at least the poorest soils. For the intermediate soils, however, in which the estimated additive genetic variance is stronger than the estimated dominance genetic variance, a form of SRS or RS for GCA would be feasible.

At present, no theory exists for determining zonations for different breeding methods. Nevertheless, it is possible to conceive of establishing environmental zones utilizing different types of gene action for a given trait. In principle, the extension of consideration for breeding multiple traits in multiple zones would not be different for this case than for using the same breeding method for every environmental zone–trait combination previously considered. It would only require recognition that in those zones where some of the traits may require hybrid breeding methods that more than one breeding population must be developed. In those zones, the selection, testing, and mating sequence would have to be considered as in Chapter 5. Otherwise, no other adjustments in the master breeding plan are required.

In the initial generations of tree breeding, we have had to ignore the possibilities of establishing optimal breeding zones and have had to focus on a general mean performance for a vaguely and intuitively understood mean environment. However, with shifting environments and the possibilities of better understanding environmental response functions, the needs for accommodating populations to variable environments and the opportunities for doing so in more optimal ways becomes more apparent. The opportunities for establishing the data and material basis for such shifts are available but can be lost by unnecessarily fixing excessively wide or narrow boundaries without examination of alternatives.

7.2 Strategies for Variable Ecological and Economic Environments

The question for breeders is how to best organize breeding populations to take advantage of present or future variations in the ecological or economic environments. For those situations in which breeding is simply organized, when no ranking changes occur for any trait over any environmental variations or for any composite value function of those traits or variations in those functions, then single- or replicate-population breeding is sufficient for present needs. The "sub-lining" proposals of van Buijtenen and Lowe (1979) are assumed to be equivalent to the replicate population proposal of Baker and Curnow (1959) if each subpopulation is maintained as a separate breeding population (Burdon and Namkoong, 1983). If not, then the sub-lines may simply constitute a temporary subdivision of a single population and not a long-term breeding population structure in the sense of Chapter 3 (Fig. 3.1).

Single- or replicate-population breeding is also sufficient for future needs if the current needs for breeding techniques and selection objectives are homogeneous and there is little doubt that future needs will remain the same. However, if future uses are expected to be more diverse than any present homogeneity, or if future needs are expected to diverge from present needs and may carry either homogeneous or heterogeneous sets of requirements, then there is a conceptual increase in the variance of the environmental factors for which we must adapt the breeding population. We can thus conceive of uncertainties about future needs as an increase in the variance of future requirements for adaptability in much the same way as represented by an actual increase in the diversity of future needs. In either case, the tree breeder who presently must adapt the breeding population to wide and relatively weakly controllable environmental variations, must consider the magnitude of changes in future generations necessary to accommodate changes in both the ecological and economic environments over long generation intervals. The problem for the breeder who wishes to effectively breed for long- and short-term gains while deploying populations with only a simple breeding system is that selection objectives may be so generalized that gain is sacrificed for simplicity. On the other hand, if breeders increase the selection intensity while maintaining the same diverse objectives, then individual breeding populations would become too small to achieve recurrent gain.

If no confidence can be placed in predicting future needs, one minimum approach is to forego cumulative gains from recurrent selection programs and to select in each generation for one-generation gains in whatever is then currently desired. A large, unimproved base population would be needed to provide the genotypic pool, but would not be selected in any directed way. This might be called an "unimproved base" population breeding system.

However, this is obviously not the only approach that can be taken to breeding for an uncertain future. It is possible to categorize the uncertainties. First, if we grant that there will always be a certain fraction of unknown and variable traits to select for, such as new resistances, ecological adaptabilities, or economically important traits, we will always require sufficiently large populations to allow for allelic variations to exist— at least in the total collection of breeding populations. Without such provisions we will have to resort to backcrossing or more mechanical and unrealistic means for gene introductions for forest trees. The cost of maintaining large populations is, of course, the lowered selection efficiency.

Second, if breeders can determine traits that will affect future values, but cannot predict the directions in which to breed for trait combinations, then the utility of future populations can be increased by increasing the variation among populations. Extremal sets of trait values such as high and low specific gravity, high and low juvenile growth rate, etc., could be selected in separate populations. With different kinds of future forest economic scenarios, an index type of selection system could be generated for each of several possible futures and a multiple set of indices developed to diversify populations according to an economically derived value variable. This is essentially the multiple-index selection strategy (MISS) proposed by Namkoong (1976). Alternatively, without consideration of economic variables, the breeder may choose several means of diversifying trait combinations such as "cornering the boxes," factorializing all trait combinations, or forming other partial factorializations of the trait combinations. An extreme form of maximizing the span of trait combinations with just $N_e = 2$ per population suggested by Mattheiss (personal communication) is to select pairs that would increase the total convex hull of the selected population. However, this was intended for just one-generation breeding and not for recurrent selection. Nevertheless, the concept might be expanded to include such a "maximal volume" concept for multiple-population breeding for recurrent selection programs.

Third, with some confidence in the future directions of breeding, we can conclude that certain of the above extremal populations would be so unlikely to ever be useful that they could be safely deleted from the program. With somewhat greater confidence, it is conceivable that a probability distribution on the values of future directions could be derived and that we might wish to use several populations more closely directed to expected future directions, much as we can optimize a distribution of populations near environmental means (Chapter 6).

Assuming that we have some information on the levels of uncertainties with respect to the traits of breeding interest, we can develop a multiple set of diverse populations from which the breeder may choose to multiply for any current, local forest management unit (compartment) that would seem desirable. Within a multiple-population system, traits may be subdivided according to the uncertainty of their future breeding objectives, and at least two programs can be developed. A first subdivision can be made on the basis of the highly uncertain traits or environments, and extremal (e.g., high–low) selection populations made for these. Within each, with presumably less uncertainty, a small set of diverse populations can be bred for the remaining uncertainty in economic or ecologic prospects for the traits that are thereby effected. Alternatively, a first subdivision can be made on the basis of the more certain traits or environments, and for each of these, more extremal (as well as median) conditions bred for in those traits of high uncertainty. The selection pressure on traits is likely to differ between the two schema, and at this time it is not clear which system might prove superior or more practical to apply.

A further possibility is to construct a hierarchy of traits according to the certainty of their evaluation. With a hierarchy of populations such as suggested by Kannenberg (1984), the large base populations might be selected only for traits of greatest long-term certainty, and the progressively smaller, more highly selected populations selected for traits of less certain long-term value. Such a plan would involve a base population with mild selection for most traits to which breeders can revert for introducing germ plasm into advanced populations. Because this would itself require either long breeding sequences for widely disparate base and advanced populations, or for a larger hierarchy with small differences between levels, it seems unlikely that this would be useful in forestry (Namkoong, 1984b).

The objective of these programs is not to maximize immediate gains but to maximize expected future values by stabilizing achievable cumulative gains and providing resilience in value to variations in the ecological-economic environment. The approaches to the problem of future versus present variability are the same; for example, if single populations can uniformly well accommodate all variations, then breeding is simplified and the standard breeding procedures described in Chapter 3 would suffice. To the extent that they cannot effectively manage the breeding populations because of different gene actions, environmental, or trait variations, then multiple populations will often be necessary. There is, however, no necessary connection between the two time perspectives. It is possible that single populations would suffice for current needs but not for the future, that multiple populations needed now may not be needed for the future, or that presently needed multiplicities do not match the multiplicities deemed necessary to accommodate future requirements.

Because we are introducing more variability into our considerations of future needs, it seems unlikely that the second case would often exist. There is more reason to expect that mismatches between present and future needs may exist and hence that the array of multiple-population objectives would differ and require that future breeding populations be developed, which would not be derived from any of the present breeding populations. Nevertheless, it seems that present breeding objectives would be closely related to most future objectives, because there is usually some continuity in the ecological and economic shifts that can occur for forests and forest products. Although there may be a wide shift, it would seem reasonable to expect that at least the extrema of current needs would be a basis for population development for future needs. Hence, the current diversity of populations can generally be expected to serve as reasonable bases for the development of the array of populations needed for future breeding. Therefore, the possible interdependence of present and future breeding needs can be accommodated within a program of multiple-population breeding.

We therefore envision that the development of future population arrays will best proceed from a foundation of diverse multiple populations as developed for the diversity of current needs. Constant revisions of the diversity of future arrays would be expected as the biology and genetics of response functions and environmental management become better known and as economic variations become more clearly discerned. In its simplest form, a whole species might be developed with a single breeding population for a given environment, with no further changes required that could not be accommodated by small changes in that single population. This is conceivable for some species having little economic significance for any purposes, with restricted ecological range and uniform management. For many other species, some further development may be reasonably expected to be desirable.

7.3 Multiple-Species Strategies

For any single forestry organization—a private company, a cooperative, or a govern-ment agency—concerns usually extend beyond single-species, uniform-site, single-objective forest management. In particular, if ecological or economic requirements are so different that the current extent of species adaptabilities are too limited for any one species to accommodate all needs, then the subdivision of effort must include multiple populations of different species. The breeding or gene management of each might well be different, and the zonations may not coincide, and hence the problem of defining optimal populational mosaics is merely elevated to the species level.

In such a program, some species may be expected to be of major importance for most of the forest management compartments, while some species of minor general importance may be of importance on certain sites or for limited commercial objec-tives. With one or a very few species in each category, it seems feasible to consider species merely as a particularly distinct division among the multiple populations being bred. In such cases, the ecological zonation as optimized for the frequency distribution of environments and the response functions of the genotypes would be more difficult to define, because interspecies differences in responses to ecological variables are likely to be more different than within-species differences. Further, gene actions and breeding methods may be different, and the trait objectives may be so widely different that several species may actually represent competing econo-mic interests in forest land use.

Then, the feasibility of developing a single strategy for population deployment as discussed in Section 7.2 is diminished. Obviously, it is not then possible to find a coincidence of ecological, economic (trait), and breeding method subdivisions. It may also not be possible to factorialize among the trait and ecological adaptability factors because no single objective function can be defined. The tree breeder can, however, provide different population sets for the major competing interest groups. For those sites and traits for which interests are common, the same populations can serve multiple interests. Where they are not common, and require different manage-ment decisions, they might still be served by the breeder who can provide different sets of populations.

Thus, while very complicated socioeconomic evaluations may make breeding more complicated, multiple-population breeding can provide the best that gene management has to offer without being bound into single strategies or value functions. The whole species program is not necessarily committed to single objective functions, nor to single private or governmental objectives.

In many cases, it is expected that single organizations will have fairly obvious species choices, site constraints, and trait objectives. In most cases, one or a very few primary species will be the primary focus of the breeding effort and will be optimally developed in a series of multiple populations. With such a "natural" subdivision, the secondary species can then be expected to have more narrowly focused management objectives to fulfill and hence can more easily be served by simple programs. At a minimum, a single population with SRS, with $N_e \leq 50$ or with a shorter expected time of usefulness, $N_e \leq 20$ might serve. For some of the secondary species that might serve some future but uncertain needs, the small breeding populations might be supported either by supplementary multiple populations or by a larger reserve (hierarchical) base breeding population.

If such a two-tiered system is adopted, some problems will exist if it is actually necessary to elevate a secondary species to some higher level of intense gene management. In such cases, the multiplicity of populations or the size of the base hierarchy population is important to maintain for future, if not present, needs. Obviously, if there are some expectations that some of the secondary species will in fact require more intensive gene management in the future, then a third category of management can be instituted. Such species programs may involve mild selection in large populations, or extremal selection in multiple, small populations that would form the basis for enhancement programs for future use.

In such cases, future usefulness and not present value is important and hence experimentation becomes a more significant objective for the breeder. Thus, in designing breeding populations, estimation experiments, and testing programs for these species, the estimation of response functions, potential adaptability, and genetic variances can be more important than progeny testing for immediate gain. In fact, in many cases in which forest management is becoming more intense, information for planning purposes is more important than the use of information for immediate gain purposes. Because forestry operations generally require long periods of time, the information gained may be only marginally useful in the current generation. However, if experimentation is directed to developing zonations, trait heritabilities, and breeding methods for subsequent generations, then the information can be available in the generation of its intended use.

Thus, for many programs, the initial breeding generations will necessarily be based on first guesses of operational feasibility on the best, but often vague information then available, and initial designations of species categories, zonation, etc. will appropriately be considered first approximations. The breeder could then plan to make iterative adjustments to those initial guesses and decide what kinds of population structures, zonations, etc. would allow the populations to adjust to expected future needs. The breeder can then decide what information will be needed to make those adjustments and to create the experiments needed for that informational base. Thus, instead of focusing on structures of the breeding program, or any current generation, the breeder would better serve forestry needs by constructing a developmental program for the continuing evolution of the forest tree populations.

CHAPTER 8

Provenance Testing, Ecogeographic Surveys, and Conservation

If we consider tree breeding in a broad sense to be a system for genetic management in which components interact with each other and respond to ecological and economical inputs, it is useful to consider broader aspects of the ways in which management system evolved. To do this, some consideration must be given to the ways that the populations at hand can be developed and to the nature of the populations themselves. In this chapter we sketch in the elements of the broad field of study and work involved in "pre-breeding" and genetic conservation as they relate to the more intensive breeding discussed in the first seven chapters of this book.

8.1 Species/Population Introduction and Testing

In agricultural practices, the introduction of new species and varieties has been difficult for advanced crop breeding and, while economically significant, has had relatively small effect in terms of the number of genes introduced. In the development of advanced varieties, the uses of wild populations or unimproved varieties has had very limited success (Cox et al., in press). The selection, backcrossing, and hybridization efforts have required many generations of crossing, testing, and selection to overcome the economic and ecological problems of adaptation (Frey et al., 1984; Goodman, 1984). Thus, the progress of materials from their initial collection through varietal enhancement to advanced breeding in the traditional hierarchical population structures has been long. One problem has been the relatively advanced performance level of the current agricultural varieties, but another has been the relative uniformity requirement of the final crop which the new varieties must fulfill.

In forestry, the gap between wild and domesticated varieties is small because tree breeding has had a short history. Furthermore, with long generation times and variable silviculture and markets for which to breed, there is a need for variation in the trees within a commercial forest generation. Therefore, it is not expected that forestry developments of multiple species and varieties would necessarily follow the commercial model of agriculture, and the introduction of new species and varieties in pure or hybrid combination may remain prominent features of continued breeding programs.

For some breeding programs, such as coastal Douglas-fir from Oregon to British Columbia, determining the primary (or exclusive) population of choice can, a priori, focus on a single natural population on the Pacific coast. At the other extreme, in many areas of east and northern Africa, and in New Zealand, the autochthonous species or varieties are not amenable to intensive or commercial forestry, and a whole new set of introductions is needed for a variety of sites and uses. For these programs, whole new forest ecosystems may have to be started with little prior experience, even for the potentially primary species. Between the two extremes, it often occurs that tests are needed to determine a species of primary economic importance, and even the choice of species of secondary interest requires some testing, research, and development. For these reasons population selection is often a significant first step in breeding programs.

Introducing populations and testing is useful for maximizing economic gain from tree breeding. However, planned introduction is necessary for germplasm conservation as well. Although breeding is considered to be a form of conservation, and the two activities might coincide at times, conservation programs in general include many more criteria to be considered in selection. Introduction for economic gain should also further distinguish between short-term and long-term economic gain. Because limited resources are available for breeding, it is likely that trees collected by contemporary breeders will represent the major portion of the gene pool available in the future. Introduction of populations designed to maximize immediate economic gain may not provide the best set of materials to be used in the future. Therefore, in introducing new species or populations, breeders need to consider three distinct general objectives: short-term and long-term economic gain and conservation. Most current tree improvement efforts around the world emphasize short-term breeding, but even for those purposes we have seen that the development of multiple-generation breeding plans often call for multiple-origin populations. These issues and techniques as related to conservation are discussed in Section 8.4.

Regardless of the needs for population introduction, managers are usually equipped with little information about the structure of new introductions, yet must simultaneously begin to develop a structure for future populations which may or may not resemble that of present populations. Knowledge of the causes of genetic variance in the natural populations and of population structure would be useful for selecting genotypes for the breeding population. This knowledge will also help address future questions of how the breeding population might be structured. Therefore, although we may wish to drastically change the breeding structure of the future populations, it would behoove us to know what potentials exist for special adaptations or problems that may be created by changing the genetic architecture.

For many forest tree species, there generally seem to be high levels of genetic variability as compared with herbaceous inbred species or those which have undergone many generations of domestication (Hamrick, 1983). The observation that forest trees in general have abundant genetic variability, however, does not explain the causes of the genetic variability that can be used to develop breeding alternative strategies. For example, some of the variation may result from a relatively high number of deleterious mutant loci held at low frequencies by mutation–selection balance (Bishir and Namkoong, 1987; Namkoong and Bishir, 1987). However, some of the observed variation may reflect some selective significance. In cases in which heterozygote frequencies at isozyme loci exceed those expected from the average allele frequency functions of Hardy–Weinberg expectations, it is often inferred that heterozygote superiority is the causal agent (Bush et al., 1987).

We discuss this issue further in Section 8.4. For now, however, note that by determining the structure of natural populations, we can infer the effects of potential causes of observed genetic variabilities and distributions. This analysis will, then, become useful in developing sampling and breeding strategies. For example, if a species contains no subpopulational structures in the wild, and if the existent genetic variation is uniformly spread throughout the species, then it does not matter where the initial population originated because any sample of equivalent size would contain all of the variation that any other sample would have. Questions would still exist with respect to whether the allelic distribution has been always like that, and whether special future adaptabilities might exist (or be developed) for different breeding objectives, even if the initial sample could be very simply structured.

For example, the species may have ordinarily been finely substructured, and recent mating or colonization events may have overridden such structures (Namkoong, 1984a) obscuring important natural variations. Even if a homogeneous structure normally exists for ecological and economic variability, it may be desirable to diversify populations for future uses. In contrast, for single-objective breeding, it may be desirable to homogenize formerly differentiated populations or to choose single subpopulations for breeding from a differentiated set. Thus, only if the natural population structure is homogeneous and the breeding objectives are homogeneous can we ignore the natural structure of populations with impunity.

8.2 Provenance Trials

When the species of choice is, a priori, perceived to be a common, locally adapted variety (or species) in tree breeding programs, there are few incentives to studying population variation, and future uses are assumed to be emergent from the initial populations. This might indeed be the ultimate position we would wish to be in, having knowledge of adaptabilities of various possible breeding populations, all of which have been suited to some economic or ecological objectives. However, at the beginning of most programs, some uncertainty exists with respect to the populations and species that might be best to develop. Even for breeding programs developed with domesticated varieties, there may exist new needs in the future, or breakdowns may occur in present adaptabilities that require shifting to substitute populations. In these cases and in cases in which new programs with allochthonous populations are required, the problem is to determine how the populations can respond to environmental variations and which traits are amenable to breeding.

8.2.1 Retrospective Testing

For programs in which populations have already been established for breeding, the main purpose of any provenance testing is to estimate the probability that any better genes or populations can be found for at least one trait, such that it can serve as a substitute or as another constituent of a multiple-population set. For particular planting sites, the question is whether there exist physical environmental or geographical coordinates of origin by which performance can be observed to vary so that resampling from particular areas or populations could be expected to provide the breeder with additional gain potential. Then, the independent variables to examine for trait responses are the environmental coordinates of the sampled populations. If it is discovered by such survey and testing that all traits are randomly distributed with respect to useful variables, then the breeder must ask what the probability is that a useful population can be found by random testing of multiple origins. The previously established populations may also be compared for their location in the distribution of possible populations, and the effectiveness of finding and using a substitute or extra population determined.

On the other hand, if such ecogeographical survey tests uncover a correlation between some environmental variables of the population origins and performance of some traits, then the total variance in yield is, to some extent, related to measurable variables and the probabilities of locating useful populations is increased. The breeder must still decide if it would be efficient to actually locate and develop such potentially useful populations.

For these purposes, multiple regression estimation procedures are required for a sampling of planting sites. When several intercorrelated yield variates determine the values of a provenance and each variate is affected by the same environmental variables but in different ways, the breeder must parameterize these relations in order to select among provenances. This parameterization is essentially a form of multiple regression extended to several traits simultaneously. For example, height growth and diameter growth are simultaneously affected by some independent variables, both are often measured, and both are often affected by genetic or soil factors in generally similar ways. Of course, the dependent yield variates are not exactly correlated in their responses, and our interest therefore centers on the pattern and strength of the joint responses.

The greater the correlation among the yield variates, the simpler the problem becomes because the results in several variates can be predicted with increasing accuracy on the basis of the behavior of any one variate. In such cases, a single (or very few) functional relationships among variates would reduce our problem to univariate analysis, which we can choose as either a linear function of all the variates or a single convenient variate for regression analysis and predict the behavior of all other variates by that function.

Once yield variates are reduced to a minimum and are expressed as functions of the independent variables, then the remaining problem for the breeder is that of determining a value function among the yield variates. This determination can be an additional critical problem if the values of the variates do not assume a linear form. Consider, for example, a curvilinear relationship among traits for individual trees. Suppose that at low stem-growth rates, fruit yield is low, but that fruit yield increases with increased growth vigor up to a point beyond which increased growth rates are made at the expense of a decline in fruiting. If managed for the dual yield of stem and fruit, there can be a problem in deciding which combination is economically best. The answer can vary widely, depending on the relative value placed on the two traits, and the nature of irregular value functions. For simpler linear models, however, extensive theory and methodologies have been developed.

In this section, we discuss the utility of some multivariate techniques in analyzing provenance testing situations in which the environmental variables influence one or more dependent yield variates. The serious investigator would profit by a study of multivariate distribution theory (Anderson, 1958) and analytical methods, as detailed in several texts such as Blackith and Reyment (1971). To briefly familiarize the reader with some of the concepts, a few of the multivariate analogs of elementary univariate statistics are summarized in Box 8.1. One use of multivariate techniques is to reduce the number of yield variates that must be measured. The same techniques may be applied to the multiple environmental variables for elimination of variate redundances (Kendall and Stuart, 1963; 1966).

Box 8.1 Multivariate Analogs of Elementary Univariate Statistics

Probability Density Function:
  Univariate normal distribution is

$$\frac{1}{\sigma(2\pi)^{1/2}} \exp[-1/2(X - \mu)^2/\sigma^2],$$

where μ and σ^2 represent mean and variance, respectively.

p-variate multivariate normal distribution is

$$\frac{1}{|\Sigma|^{1/2}(2\pi)^{p/2}} \exp\{-1/2[\underline{X - \mu}]'\Sigma^{-1}[\underline{X - \mu}]\}$$

where Σ represents the variance–covariance matrix of the p variables, and

$$[\underline{X - \mu}] = [X_1 - \mu_1, X_2 - \mu_2, \ldots, X_p - \mu_p]'.$$

Regression:
  Single dependent variable, y, and single independent variable, x.

$$\text{Mean: } \bar{Y} + (\sigma_{yx}/\sigma_x^2)(X - \mu_x)$$

  Residual variance: $\sigma_y^2 - (\sigma_{xy})^2/\sigma_x^2 = \sigma_y^2(1 - r^2)$, where $r = \sigma_{xy}/\sqrt{\sigma_x^2\sigma_y^2}$

  Multiple x variables.

$$\text{Mean: } \underline{y} + \underline{\sigma}_{yx}\Sigma_{xx}^{-1}[\underline{X - \mu}]$$

  Residual variance: $\sigma_y^2 - [\sigma_{yx}]'\Sigma_{xx}^{-1}[\sigma_{yx}]$
                where Σ_{xx} is the covariance matrix among the x's.
                σ_{xy} is the covariance vector between y and the x's.

  Multivariate regression with several x variables and y variates.

$$\text{Mean vector: } \underline{\mu}_y + \Sigma_{xy}\Sigma_{xx}^{-1}[\underline{x - \mu}]$$

  Residual covariance matrix: $\Sigma_{xy} - \Sigma_{xy}\Sigma_{xx}^{-1}\Sigma_{xy}$
                where Σ_{yy} is the covariance matrix of y's and
                Σ_{xy} is the covariance matrix of x's and y's.

Simple tests:
  Univariate t tests:

$$t = [\sqrt{N}(\bar{x} - \sigma)]/\sigma, \text{ where } t^2 \text{ is distributed as an } F_{(1, N-1)\text{df}}.$$

  Multivariate tests:

$$T^2 = N[\underline{x - \mu}]\Sigma^{-1}[\underline{x - \mu}], \text{ where } T^2(N - p)/(N - 1)p \sim F_{(p, N-p)\text{df}}.$$

APPLICATION OF PRINCIPAL COMPONENT ANALYSIS

The reduction can be accomplished by component analysis rather than by multiple regression. If, to start with, independent p variables may account for part of the genetic variance, the problem is to find whether collinearities exist among any subset of the variables. A collinearity is said to exist between two variables when the occurrence of one variable at a specific level fully determines the other. A single linear relationship that completely describes the joint variation can then substitute for the two original variables, or, conversely, one of the variables is redundant. If there are three variables and one is fully dependent on the other two, a collinearity in three-dimensional space exists and all the variation is in a two-dimensional plane. Then, a single collinearity is said to exist and the rank of the space is two. In component analysis, a series of lines (in the original p dimensions) is successively and orthogonally fit to reduce the residual variance about the lines. These lines are the principal component vectors. If a single line describes all of the variation in the variables, the first component would be that line.

If fewer than p one-dimensional transforms are required to account for almost all of the variance, then there are, perforce, linear dependencies among some of the p original variables. Dependencies imply that some of the variates that have been measured can be fully explained or replaced by a linear function of other variates. Therefore, the removal of at least one variable is possible. Of the p orthogonal vectors, the one corresponding to the smallest root of the standardized covariance matrix (i.e., the correlation matrix) presumably represents almost a random vector in the orthogonal residual space. The original variate which has the largest coefficient in this vector is that which can presumably be best explained by the others, and hence is a likely candidate for discarding. Using the remaining original variates the estimation of further components may then be repeated for discarding variates for as long as zero or near zero roots continue to exist.

CANONICAL CORRELATION ANALYSIS

The principal component analysis procedure for reducing dependent variates can also be applied to reduce the number of independent variables (Kendall, 1961). Like other regression techniques that reduce the variate space, this method is subject to the usual restrictions on interpretation of cause and effect or anything other than simple association. Although the component analysis possesses some advantages over the more common stepwise procedures, it is scale dependent. The standard procedures for variable reduction in multiple regression may therefore be more useful parts of an analytical system. When interest exists in finding the linear function of independent variables (X) that can best fit a linear function of dependent variates (Y), the regression coefficients for the X variables and the component coefficients for the Y variates can be simultaneously chosen to minimize the error variance of the principal component. The technique is known as canonical correlation analysis, and although not yet used extensively in forestry, it is potentially useful in provenance analyses where several traits vary simultaneously in response to several site variables and the greatest degree of explainable variation is desired.

Nonlinear Regression and Response Surface

If consideration is restricted to a reasonable number of dependent yield variates and independent environmental variables, the problem of parameterizing the joint relations is often simply one of describing the regression effects. In the univariate case, a response surface is estimable and the combination of source-environment variables that gives the greatest expected reduction in residual error in the dependent yield variate is identifiable. If a simple quadratic surface is hypothesized and estimated, then optimum levels of the X environmental variables that would give a maximum yield can be estimated. For example, the dependent Y variable may be some measure of growth and the independent X variable may be fertilizer level or the latitude of the seed source, both of which may have some intermediate optimum level for maximum Y. The error in estimating the optimum point for maximum Y is also estimable. If a simple quadratic response line of yield to a single environmental variable, for example, latitude, were to be estimated by $Y = \hat{b}_0 + \hat{b}_1 X + \hat{b}_2 X^2$, then the maximum likelihood estimate of the X corresponding to a maximum Y is $-1/2\,(\hat{b}_1/\hat{b}_2)$.

The standard error of the maximum Y can be estimated by such procedures as given by Kendall and Stuart (1963, chapter 10, p. 232) and the confidence belt estimated as that which would be appropriate for regressions derived from a normal distribution of errors of Y. Regions in which the environmental variables are of acceptable levels may then be set up by simple graphical or more sophisticated techniques. We may extend the number of different environmental variables to two or more, but remain in the univariate case and describe a quadratic response surface, for example, by

$$\hat{Y} = \hat{b}_{00} + \hat{b}_{10}X_1 + \hat{b}_{20}X_1^2 + \hat{b}_{01}X_2 + \hat{b}_{02}X_2^2 + \hat{b}_{11}X_1X_2$$

The levels of X_1 and X_2 giving maximum Y are

$$\begin{bmatrix} X_1 \\ X_2 \end{bmatrix} = \begin{bmatrix} 2\hat{b}_{20} & \hat{b}_{11} \\ \hat{b}_{11} & 2\hat{b}_{02} \end{bmatrix}^{-1} \begin{bmatrix} \hat{b}_{10} \\ \hat{b}_{01} \end{bmatrix}$$

The estimation of the regression coefficients of the linear model is well known and can be written in terms of estimating the vector β by $(X'X)^{-1}X'Y$ where X is the matrix of the levels of the sampled variables and Y is the vector of the yield variate at each level of the independent variables. If the model is extended to the multivariate case, each dependent yield variate would be characterized in its response to the environmental variables by its specific vector of regression coefficients. The estimation is a simple extension of the usual univariate procedures for estimating a β matrix instead of a vector β. For example, for the same n samples and two dependent variates Y_1 and Y_2, the model takes the usual form:

$$Y = X\beta + \varepsilon$$

with the change in the meaning of the expressions such that

$$Y = [Y_1 \quad Y_2], \qquad \beta = [\beta_1 \quad \beta_2], \qquad \text{and} \qquad \varepsilon = [\varepsilon_1 \quad \varepsilon_2]$$

Because the X matrix remains the same, X can be used as it was for univariate case. Then, the β matrix is estimated as

$$\beta = (X'X)^{-1}X'Y_1Y_2,\dots$$
$$= (X'X)^{-1}(X')(Y)$$

The covariance between the regression coefficients is estimated by $[X,X]^{-1}S_{ij}$, where S_{ij} is the residual covariance between traits Y_i and Y_j.

An Example of Response Surface

The data of Wells and Wakeley (1966) on the performance of loblolly pine of various seed sources in a plantation in Dooly County, Georgia, will serve as an example. The independent X variables are the January minimum temperature (X_1) and summer rainfall (X_2) of the source locations, and the Y variates are survival (Y_1) and height (Y_2). The regressions are drawn in Fig. 8.1. The regression equations estimated on the nine source locations are

$$\hat{Y}_1 = b_{i00}b_{i10}X_1 + b_{i20}X_1^2 + b_{i01}X_2 + b_{i02}X_2^2 + b_{i12}X_1X_2$$
$$\hat{Y}_1 = 62.9 + 2.91X_1 - 0.06X_1^2 - 1.98X_2 - 0.09X_2^2 + 0.09X_1X_2$$
$$\hat{Y}_2 = 135.5 - 6.79X_1 + 0.11X_1^2 + 0.59X_2 + 0.09X_2^2 - 0.08X_1X_2$$

The conclusion of Wells and Wakeley (1966)—that the fastest growing trees in the Dooly County, Georgia, plantation were from regions with warm winters and wet

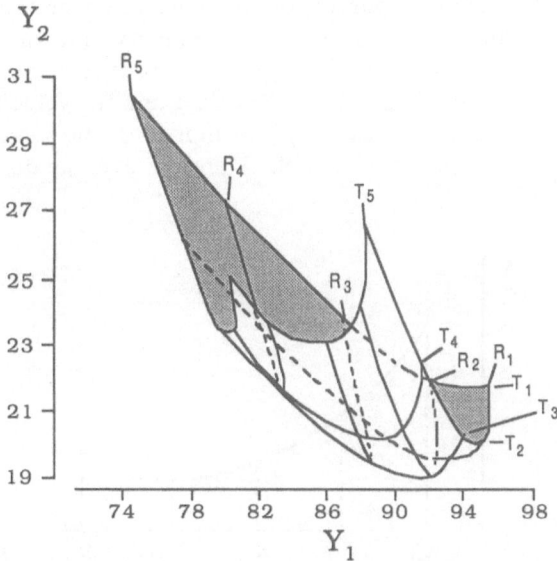

Fig. 8.1. Quadratic response surface of the simultaneous effects of rainfall and temperature on survival and height growth.

summers—is generally supported by the present analysis. The quadratic surface model may not be the best fit for the data, but it is evident that warm winter at the seed source favors growth but not survival and that low summer rainfall at the seed source favors survival but not growth (Fig. 8.1). The sampling of source environments in this plantation is not sufficient for more precise source evaluations. It is generally not wise to extrapolate from a model that at best may mimic a set of data and to try to discern cause and effect, particularly when such high residual errors exist. However, the generally negative correlation between height and survival for both the total variables and the residuals after regression indicates that selection for both would result in opposing selection forces. The deviations from regression allow for some combined selection, and the data suggest that deviations exist in the direction of intermediate to high winter temperature and intermediate to low summer rainfall. If selected, these sources might best maintain both height and survival without great loss in either. Provenance hybridization to combine genes for both may be warranted. If survival is of less importance in this range, the desirable vector of selection would more heavily favor the warm winter–wet summer sources.

VALUE FUNCTION

A problem for the breeder in provenance selection is how best to find a region for seed source sampling or its unique set of environmental conditions that simultaneously optimally affect the important yield variates. If we establish a space defined by these dependent variates, it would be possible to locate a point representing the vector of the several dependent variates, denoted by Y corresponding to each set of environmental conditions, denoted by a vector X. Then changes in X over the independent variables define a surface of changing values in the Y variates. The problem then is to define some joint evaluation of all Y variates and then to find the maximum value of that joint value function. An optimum X may not be unique, however, depending on the shape of this Y surface and the value function used.

Geometrically, the simple regression problem is to define some functional relationship between an independent variable X and an average dependent response variate Y.

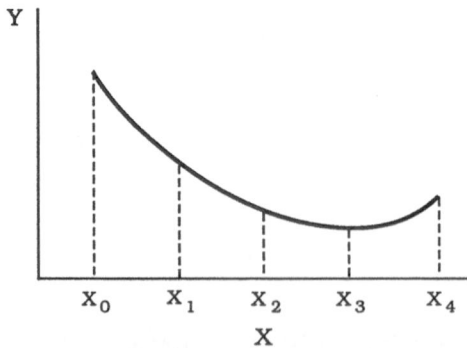

Expanding the case to more than one X in multiple regression requires description of the response surface of Y to the X's.

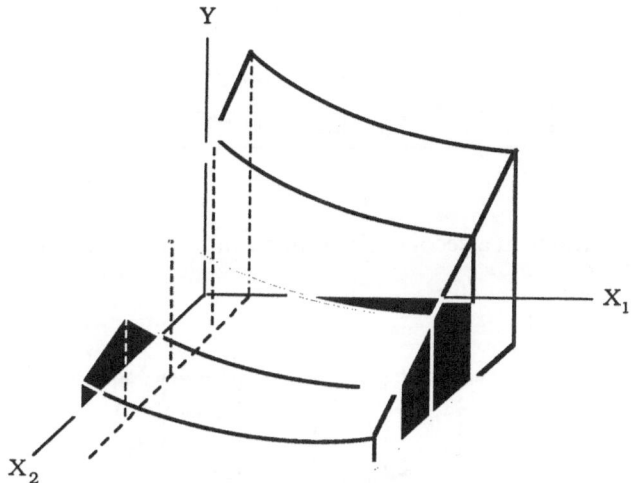

This may be projected onto the Y, X_1 plane when X_{20}, X_{21}, X_{22}, and X_{23} are the projections of Y at the various levels of X_2. Considering now that two dependent Y variables may exist for a single independent X variable, there generally is only one mean joint response point (Y_1, Y_2) for each point in X. In that case, the response of Y_1 and Y_2 to X_1 is a line in three dimensions.

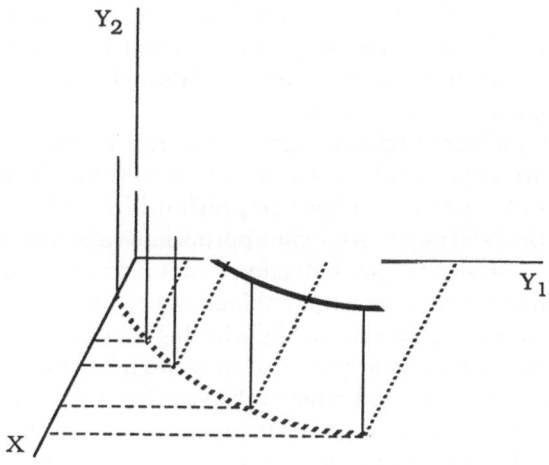

This may be projected onto the Y_2, Y_1 plane.

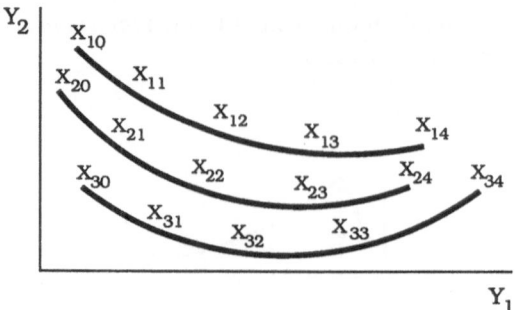

Then similar variations in a second X variable can be projected onto this plane.

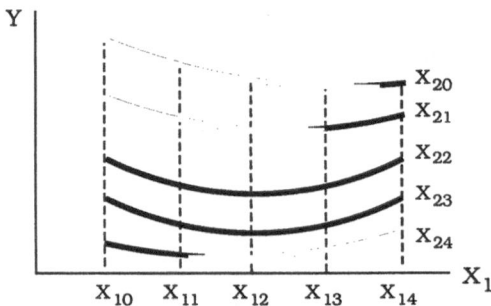

This kind of projection then graphically displays the joint response of two dependent Y variates to two independent X variables. It should be noted, however, that the proviso was stated that there is only one mean joint response point to each point in X. In fact, however, there are residual variances, in each Y variate, and some residual covariances and correlated response of Y_1 and Y_2 at each X point. In terms of provenance analyses, these are the intrapopulational variances and covariances, which may be of considerable interest.

More formally, the linear relations are describable by the equation $Y = X(\beta)$, where Y is the vector of yield values, X is the vector of the environmental variables, and β is the matrix of regression coefficients previously described. The problem is to pick the vector X that will maximize some appropriate value function of Y. Because these relations are scale dependent, reduction to a standardized or other established basis is recommended. Simple linear combinations of the variates may always be constructed, but often the gain achieved by selecting for conditions maximizing one trait will not be optimum for other traits. It may often be best to pick the environmental vector that will assure that none of the variates suffer much loss and which thus will maximize the minimum gain. Of course, it is possible to construct solutions to maximize the sum or product of all variates. For any criteria, the surface representation will allow one to examine how the alternative solutions will differ and will help to determine if intermediate or diversity selection is best.

It is illuminating to consider how different vectors might influence the choice of source environments in the loblolly pine example. An infinite number of functions could be written for the relationships between the yield variates. If equivalent and linear economic weight is given to survival in percent and height in feet, the expected value would be

$$E(Y_1 + Y_2) = 198.4 - 3.88X_1 + 0.05X_1^2 - 1.39X_2 + 0.01X_1X_2$$

Here X_1 has about twice the weighting of X_2. If height is so important that the scale of value is $Y_1 + 10Y_2$, the regression would be

$$E(Y_1 + 10Y_2) = 1.418 - 65.0X_1 + 1.04X_1^2 + 4.0X_2 + 0.81X_2^2 - 0.71X_1X_2$$

In this linear weighting of Y_1 and Y_2, the value (or objective) function is a straight line. In a multiplicative model of value, the value function would be a hyperbolic line. In the linear form, and with a $1:10$ weighting, a negative weighting is given to winter temperature (X_1) and a positive weighting to summer rainfall (X_2). Both linear cases indicate that several X_1, X_2 combinations could provide good value yields but also that there is a unique minimum. Because only unique minima exist, the analysis is useful for indicating unique sources to avoid rather than unique sources to choose. On the other hand, hybrids of the alternately good sources may be quite valuable. These regression solutions for provenance selection are simply multivariate extensions of univariate theory. Estimation problems in multivariate analysis are only slightly more involved, although distribution theory is often more complicated.

In addition to the statistical questions, chief difficulties are in deciding whether to put selection emphasis on between-provenance or within-provenance differences and in interpreting the pattern of genetic variance and covariance in terms of population structure. These methods merely facilitate consideration of the joint processes of several variates under the influence of several environmental variables. Because many problems in forestry involve the simultaneous evaluation of several traits on single trees or populations because of the multiple values that exist in forests, multivariate analyses are likely to be more important in forestry than in other agricultural sciences. Not only are multiple uses of forest lands commonly required, but over the duration of a single forest population, changing uses will be imposed by human activities. Thus, for the forester, it would be most appropriate to consider univariate analyses as a special case of multivariate analyses rather than the multivariate case as an extension of the univariate case. Regardless of the need, we have been largely limited to univariate model approaches to genetic analyses.

These kinds of provenance testing or ecogeographic survey testing might be termed retrospective in the sense that they focus on the utility of populations for predetermined areas. Testing is often done with the purpose of establishing the areas to which particular populations may be best suited. Thus, for a set of populations or genotypes, their response to environmental variables of the planting sites is the focus of interest. For these purposes, a wide array of planting sites is chosen and the response to the planting site variations is estimated. Such tests are in fact, the G × E tests discussed in Chapter 6 in which stability analysis and linear and nonlinear response function analysis form the basis for determining zonations.

8.2.2 Dual-Purpose (Prospective) Provenance Trials

It is possible to consider the retrospective trials as an examination of multiple-source variables for single planting sites independently of the prospective trials, which examine the performance of unordered genotypes on multiple planting sites. In each type of trial, the use of balanced experimental designs has often provided data such that the effects of source or site environment or yield could be examined with respect to measures of geographic, soil, or other variables. As with agronomic crops, however, the efficiency of unbalanced or partially balanced block designs has proven great enough that it is often economically feasible to carry out progeny testing or provenance experimentation only if unbalanced designs are used. In addition, because silvicultural systems rarely exclude moderate mortality, and sampling is often unbalanced to begin with, the experiments often become unbalanced even when executing planned balanced designs.

With large numbers of entries and sites to manage, the traditional approach has been to use Partially Balanced Incomplete Blocks (PBIBs), including various forms of lattice designs (Burley et al., 1966; Snyder, 1966). In these designs, large blocks are broken into smaller sets that permit treatment of source performances to be contrasted within smaller block sites with less error. The loss in precision of contrasts between entries in different blocks is compensated for by ensuring that all pairwise contrasts are included in a small block in some replicates. While these PBIB's allow for all contrasts to have comparable levels of precision, severe restrictions are placed on the number of entries allowed. Efficient sampling for testing suitable provenances on a sequence of sites would require that some changing subset of sources be tested on each site if some choice in source sampling is possible. A partial sampling design may be like this:

		Source				
		A	B	C	D	E
	a	×				
	b	×	×			
	c	×	×	×		
Site	d	×	×	×	×	
	e		×	×	×	×
	f			×	×	×
	g				×	×
	h					×

Overlapping sources among planting sites are required to determine general source effects and to distinguish between source × planting site interaction effects and general source average performances.

A complete factorial sampling of all sources on all sites would provide a complete picture of value at each combination, and we could then describe a three-dimensional factorial response surface of value:

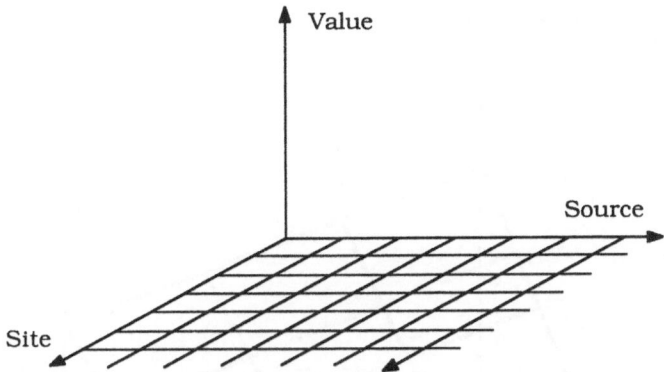

If source and sites are identically ordered and the order has some relation to value, some simple surfaces may be described for some simple hypothetical results. If sources were identical in response to all sites, the value surface would be like:

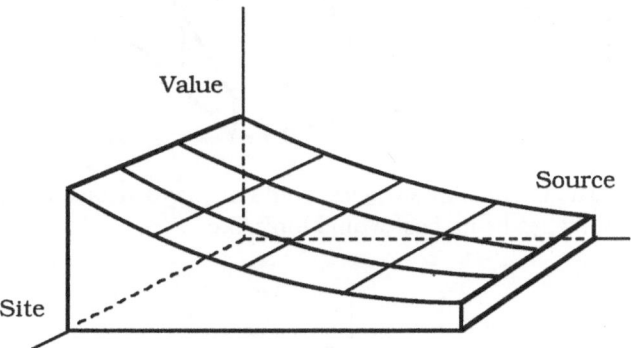

If sites differed but had identical effect on all sources, the value surface would resemble:

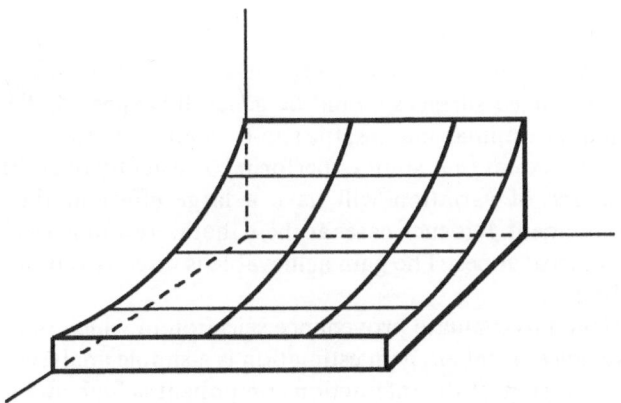

If both sites and sources differed but no interactions existed, the value surface would resemble:

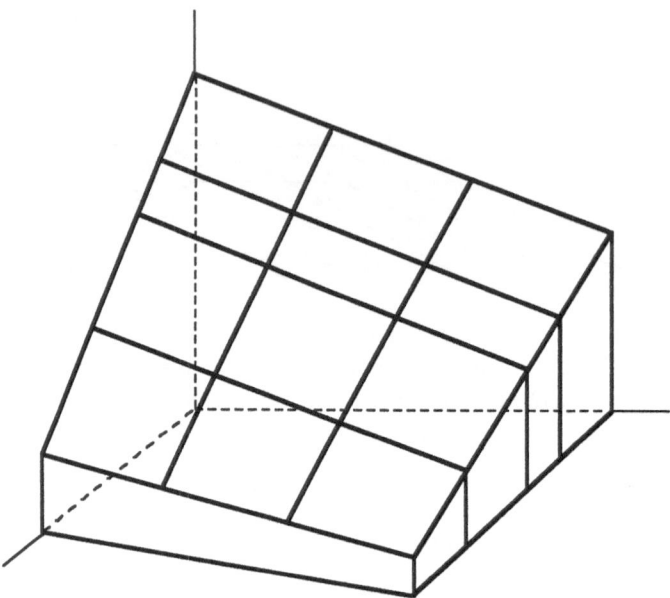

If interaction existed, irregular surfaces would display various forms. A condition of local sources always being best would look like:

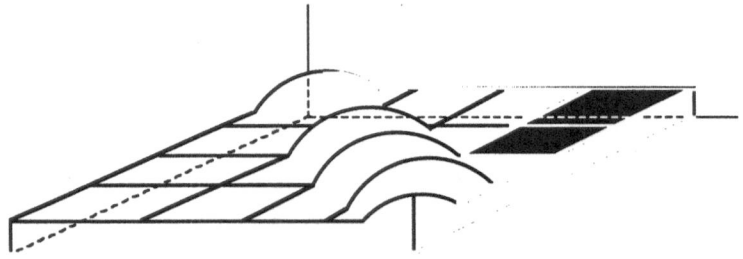

Less regular and mixed surfaces would be generally expected. To the extent that each site has unique optimal sources, the site-by-source interaction can be expected to be high. To the extent that sources perform consistently over different sites, the provenance source of variation will have a large effect at the expense of the interaction component. It is such cases as these that were alluded to in the discussion of optimal ecological zones. The gain achievable is directly estimated by the mean differences observed.

In tests designed to evaluate provenance selection in which only a sample of all possible provenances is taken, gain estimation is a simple analysis of ordinary gain estimates by regression. If the interaction component is high and selection is to be generally among the best test sources, then the expected selective advantage for starting with the best sources is the selection differential $\times h^2$ (provenance), where h^2 (provenance) is covariance (provenance test value, breeding value) divided by variance (test values), and where the selection differential is the difference between the population mean and the mean of the selected provenance. In this case, the numerator covariance of the provenance h^2 will be largely the interaction variance component plus any contributions from persistent provenance performances.

In any case, it is possible to start consideration of design for both retrospective and prospective objectives and the analyses they require. In a site-source test of *Populus deltoides* in the middle United States, for example, 10 provenances of 50 at each of 20 planting sides could be planted. In the most southeastern planting, the 10 southernmost sources were sampled, and in the next most southern planting, the two most southern sources were dropped and the two next northerly sources were added. This chain of blocks was nested to the most northern sites with the most northern sources, and because the planting site range was smaller than the sampled source range, the plantings always included local sources and both more southerly and northerly sources.

In cases in which no patterns are expected, or the source blocking is random, the variance of the mean contrasts will vary according to the number of times the sources share the same block, have some other sources in common in the same block but do not appear together, or are even more tenuously connected. The problem for the experimenter is that the variance of each contrast can differ and it is not clear if an average variance or an extreme variance are good measures of design quality (Friedman and Namkoong, 1986).

In addition to the testing and estimation objectives there is often an additional objective of using provenance tests to provide material for establishing a base breeding population. Then, not only does the test serve as a means of identifying useful populations for some traits, it becomes the selected population(s). If several source populations are selected they may serve as separate multiple populations and the individuals within each could merely be the best ones for the population objective. In that case, there may exist a family structure within sources, but individual trees within each family may be selected to serve as individual parents. In that case, plot design should include more trees per plot than might be designed for family mean estimation, and a balance must be struck between creating a large number of family plots to better estimate family responses and a low intertree contrast variance with high intraplot selection differential.

Thus, from the simplest case of single-objective breeding and single-provenance level selection to more complex multiple-objective breeding with provenance, family, and individual tree selection, provenance tests can be designed to serve the initial generation of forestation as well as to serve as supplemental populations for later breeding generations.

If a mixture of sources is selected for starting the breeding population, materials with different gene frequencies will sometimes be mixed in the population. Any dominance and epistatic effects will then produce genetic recombinations and genetic variances in the F_2 generation that are unforeseen in the parental or initial crossing generations. It would, of course, be beneficial to start into recurrent selection either for a hybrid system or for general combining ability with a hybrid base population with some experimental information on the importance of non-additive effects. However, if most provenance crossing displays additive and averaging effects and little dominance or heterosis, as appears true for most pine species, then provenance selection is simply a higher organizational form of family selection. Then, selection may be made in a tandem fashion: first provenance or source selection, then family and individual. Simultaneous selection on an individual-tree basis is also possible if some index weight is given to family and provenance collateral relatives in judging individual worth. The effects of linkage disequilibrium, however, will be felt for several generations. Other than performance testing, the ancestral structure of populations can be discerned by analysis of traits that have no selection history such as is usually claimed for most electrophoretic enzyme loci or for DNA polymorphisms detectable by DNA restriction mapping. Analyses of forest tree populations by these techniques not only reveal details of the mating structure of trees (e.g., Brotschol et al., 1986; Tigerstedt, 1984), but also possible associations with selectively important loci (Geburek et al., 1987; Bergeman and Scholz, 1987). However, such loci are usually nearly impossible to relate directly to traits of breeding importance but may nevertheless be useful to reveal ancestral structures.

8.3 Provenance Testing and Ecogeographic Surveys

In addition to sampling the planting sites that would be more commonly present in predicted future environments, it is often useful to include more extreme sites to better understand and estimate response functions. In addition to differences in family responses, the genetic variances in more extreme sites may be different than in the more usual sites. For example, in a fertilizer experiment with loblolly pine families, Roberds et al. (1976) found that families differed in the shape of their response to urea amendments on a nitrogen-deficient soil. This resulted in a high ratio of σ_D^2/σ_A^2 in the extreme environments with low average height growth and a low ratio in the moderate environments with high average growth rates. Thus, different genotype response functions can produce different rankings of genotypes over a range of environments, implying that effective gene action in particular environments are different. Although the latter estimates may be reifications of assumed fixed gene effects, inferences for selection can be drawn. Hence, the concept of environmental zonation for different populations or genotypes can be expected to sometimes imply that different breeding methods should be used in the separate zones. Hence, such breeding design questions are often best answered by designing experiments that estimate response functions well by including wider environmental variations than otherwise necessary.

Similarly, population sampling might often best be conducted by including samples from a wider range of sources than might be judged as reasonably likely to yield directly useful breeding populations. For species that contain divergent population structures as the result of either strong migration barriers or strongly divergent selection effects, wide sampling is needed to capture the alleles at all varying loci. This would be particularly true for species in which both mating demes and selection demes are segregated, because in such cases many essentially unique alleles can persist in many different populations. With wide migration of seed and pollen, only very strong and divergent selection effects or an average heterosis can maintain multiple alleles. For alleles with wide average (arithmetic or geometric) heterosis, the alleles could be well spread throughout the population, but for allelic variation protected by divergent directional selection and effective limits to migration, different alleles may exist in high frequency only in the more extreme environments. With wide seed migration and restricted pollen migration, the strength of diversifying selection need not be as strong to maintain genetic variation, and with restricted seed migration, even weak selective differences can maintain variation (Gregorius and Namkoong, 1983; Namkoong and Gregorius, 1985).

While it seems clear that tropical hardwoods, even with high outcrossing (O'Malley and Bawa, 1987), would commonly be expected to have migration that is severely restricted and selection that is perhaps highly localized by biotic factors (Hamrick and Schnabel, 1985), it is not clear whether even some temperate or even boreal zone conifers might also often be structured more finely than anticipated (Namkoong, 1984, 1985a; Tigerstedt, 1984).

Clearly, substantial variation occurs among physiological traits and morphological traits with respect to the pattern and extent of population subdivision. There are also, obviously, differences between pairs of morphological and isoenzyme marker alleles, and many different isoenzyme marker allele patterns exist. The problem remains as to which of these patterns carries significance for the breeder.

8.4 Genetic Resource Conservation

It is not the objective of this section to discuss the need for conservation. The needs and goals vary depending on nonbiological factors. Questions on the appropriateness of goals are important to discuss, but for purposes of this book it is assumed that such needs and goals are defined. Instead, we discuss strategies of developing conservation measures. Because limitations of available resources always exist, we discuss how our biological knowledge can be best put to use under such limitations.

The design of programs for conserving biological diversity depends on three things: (1) the distribution of the diversity to be conserved, (2) the amenability of the species to the different conservation strategies, and (3) the objectives of the conservation programs. Biological diversity lies not only in different geographic locations and in temporally variable patterns, but also at different levels of biotic organization. Biological diversity may be discussed in terms of variation at the molecular, subgenic level of organization, at the genic and individual organismic level, and at the population, species, and ecosystem levels; therefore, it can be measured in many ways. In addition to variations in their diversity, species differ in their adaptability to techniques of tissue culture, seed storage, or sustainability in other than natural conditions. Therefore, a complete strategy of conserving diversity must address the status of species needs, the biological appropriateness of technologies, and the uses of populations so that the efficacy of the techniques can be objectively judged.

Six types of germplasm management programs can demarcate differences among program elements: advanced breeding, enhancement, collection and evaluation, ecogeographic survey, targeted sampling, and biological reserves. This is an arbitrary subdivision of a continuum of activities used in gene conservation and breeding, but is useful for condensing discussion around real operational differences. In each of these management categories, various in situ and ex situ techniques can be considered. Included under in situ are maintaining strict reserves with natural survival and reproduction; population sampling, surveying, and evaluation in natural conditions; collection, enhancement, and possibly even breeding by such management techniques as thinning, breeding, and transfer; and constructed populations established within the species natural range. Included under ex situ are seed, pollen, and tissue storage; population trials outside the species' natural range; and population management or naturalization away from the population's original site, including those of domesticated varieties. Obviously, this too represents an arbitrary subdivision of a continuum of activities that is, nevertheless, useful for understanding the array of activities that are involved in global germplasm management. While it is not surprising that the utility of the various techniques and management strategies are viewed quite differently for specific programs, those differences should not obscure the underlying unity possible for a global strategy to conserve multiple levels of diversity with multiple techniques.

8.4.1 Levels of Biological Organization and Conservation Strategy

For some objectives, the nonextinction of species is sufficient, but for some breeding objectives genic-level diversity is necessary. Because it is a practical impossibility to conserve all genes of all species and is not necessary for all purposes in any one species, we must define the levels of biological organization and the conditions of their conservation that are necessary or sufficient for our objectives.

Recent studies on the molecular structure of genes and on levels of allelic variation within and among loci leave little doubt that a wealth of genetic variation has existed in most plant and animal species. Further, studies on the mating and selection of individuals and populations within species are revealing a more complicated view of intraspecies structures and the possible significance of evolution at the population level. Evolution at this level may also be driving and be driven by other species such as competitors, predators, or diseases, and hence may affect the structure of multiple-species ecosystems. However, although variations of lower level effects are necessary conditions to generate variations at higher levels of organization, conservation at one level is not sufficient for conservation at another level; hence, considerable debate exists on the relationship between population-level and ecosystem-level variation. Nevertheless, the fact that species generally contained diverse structures is not debated, and the potential utility of this variation is also beyond serious doubt.

At issue is whether the present extant variation is uniformly distributed throughout the species range or if it is grouped. If it is uniformly distributed, then any one sufficiently large sample would contain all variations, but if it is grouped, then multiple-population samples are required. At the one extreme, a species may exist as a collection of disparate populations with some past but little present intermating and with widely different adaptations to different local selective environments. In such species, the divergence of its constituent populations may be so great as to render the species concept almost meaningless, as each local population may be generating unique genotypic arrays. In these cases, conserving diversity requires conserving each population. At the other extreme, a distinct species may exist in a unitary habitat with complete intermating opportunities for all individuals and with no genetic substructures. In such species, population size and selection effects are not complicated by differences among populations, and even if variation among demographic groups and at different levels of genic organization make our understanding of species dynamics more difficult, the species can be treated as a single evolutionary unit. Thus, the pattern of diversity at the population level of biotic organization determines the adequacy of single versus multiple sampling strategies. Population-level conservation is sufficient for genic and species conservation if the species is very simply structured, and is necessary if very diversely structured.

Obviously, most species are at neither extreme, but rather have some restrictions in effective migration in at least one sex, exist over an array of selectively different environments that share partially common boundaries with mating–migration boundaries, and experience temporal variations in both (Roughgarden, 1979). While some species seem to develop highly specialized and integrated features of their adaptive behavior with their mating and gene action systems, other species seem to be able to change their mating–migratory behavior almost simultaneously with their changing environmental circumstances (Hedrick, 1983).

In addition to directed forces of evolution, stochastic variations within and between populations can provide further diversity. Thus, from the evidence of the relatively few cases in which genetic studies of species include anything but the most simplistic ecological effects, the formation and significance of patterns of genetic variation within species is neither uniform nor easily estimated. It may at times be a practical necessity to conserve species with only one population sample, but in general some multiplicity of samples will be desirable. If the species is known to have a center of diversity or a main population concentration with border satellite populations, a reasonable sampling strategy would focus on the main body, with extra samples spaced among the outliers. However, if the species variation is known to be uniformly dispersed, or if we are ignorant and must assume a uniform probability of sampling useful genes, then there is no central population and samples must be dispersed.

Another element of the biological nature of species that affects the relative adequacy of ex situ conservation is whether the particular in situ environment is an important feature in the survival or utility of the species populations. In particular, when the environment is considered to include other species of the biome, the potential exists for greater evolutionary complexity and for the significance of subtle genetic variations to be obscured from our understanding (Levin and Udovic, 1977; Namkoong and Selgrade, 1986).

Multiple, interspecies relations can be important in defining critical elements of both survival as well as reproductive success. However, even considering only pairwise interactions of species that contain genetic variations within each, the existence of single, globally stable equilibria is clearly not the only feasible way that strong interactions can maintain the coexistence of two species (Selgrade and Namkoong, 1984). Populations may change from one configuration of population densities and gene frequencies to another, or may cycle among such states. Because such behavior can be driven by a constant set of selection coefficients, the effect of interspecies relations can be different from place to place and time to time. This makes the statistical analysis of evolution very difficult, but, more importantly, it implies that different population samples can be driven to different states of genetic and species composition by the same selective forces.

Thus, even without the existence of other environmental differences among populations, natural evolutionary processes can produce a diversity of multiple populations, all of which may be significant in the evolution of each constitutent species and hence would be desirable to sample. Moreover, this would imply the necessity of not only multiple sampling within each species, but may also imply the necessity of in situ conservation to ensure continuing the coevolution necessary for species survival and evolution.

Thus, the biological nature of diversity may determine the desirability of different in situ and ex situ techniques, but the risks and costs of those techniques may constrain the final program choice. Some species may be physiologically incapable of maintaining seed or pollen viability, or growth in tissue culture, or even of enduring stand management techniques. For some other species, such techniques may be possible but costly, or may require more knowledge than we have to ensure the security of any collections. On the other hand, the security of in situ conservation areas may be in political or economic jeopardy or may require locations so remote from use that they are inaccessible for practical use. When physiologically possible, ex situ techniques may be the only way to conserve genes and species even if this isolates samples from beneficial evolution. Thus, the biological and geographical amenability of species to currently available conservation techniques can determine their practical applicability.

8.4.2 Conservation Objectives

The effectiveness of techniques will substantially depend on the objectives of the conservation program. For the immediate use of commercial breeders, long-term evolutionary potential may be relatively unimportant and any sample of immediately useful genotypes is sufficient. However, the existence of diverse evolutionary paths may be important for the multiple environmental services that the biota perform and from which humans benefit. These in situ services include cryptic effects on our physical environment, such as water and air regimes, and secondary biological support of agricultural systems such as pest and pathogen control and nutrient recycling and waste absorption. Other services, such as the direct effects of supplying new biological products for medicine, food species, or the provision of genes that might be transferrable into breeding systems, can still be provided by in situ programs. However, the breeder who is trying to introduce modifications to well established varieties is primarily interested in particular genes or traits and not in general trait clusters or adaptabilities. For these purposes, saving genes with efficiency, security, and completeness are the main objectives for immediate use of biological diversity, and ex situ, if possible, conserves genes more conveniently.

For cases in which more substantial changes are either necessary or desirable, then whole-population breeding using reserve populations for substitution of other varieties or for hybridization may be necessary. In these cases, alternate populations may be developed for adaptation to particular environments or pathogens, or for economic requirements, and these may be used simultaneously or in tandem in multiple varietal combinations. They may be targeted for particular sets of environments (Comstock, 1977) or simply designed for adaptation to an array of environments along some measurable environmental gradient. For these cases in particular, the adaptations which species develop for different environmental conditions in situ may well be useful to further enhance by breeding for diverse objectives. There is, therefore, a wide array of genetic management programs appropriate for the different objectives of conservation and/or the different stages of development required for use.

8.4.3 Commercial Species

For species that have some commercial significance, the actual variety used is but the last population in a sequence of increasingly selected populations. To achieve a successful commercial culmination, activities can be considered to begin with surveys of extant variability. An ecogeographic survey may be an informal observation of the origin of species variants, but may involve more elaborate in situ studies or may be coupled with ex situ tests relating performance to variations in the sites of origin. Such field surveys are proposed for wheat in Turkey and the Near East. This is followed by selective breeding to further enhance the value or adaptability of the population for existing standards in the areas of use. Such "enhancement" populations may involve several generations of selection, crossing, and backcrossing. For example, oats and sorghum breeding with wild relatives takes four to six generations of intercrossing before new varieties enter advanced breeding and eventual commercial release (Frey et al., 1984). Alternatively, populations of independent origin can be selected for adaptabilities to particular environments over several generations of recurrent selection (Goodman, 1984).

Tracing this sequence backward in time, the breeders' base populations or foundation stock are defined as those populations from which ordinary selective breeding can develop varieties of direct value to commerce. These are the populations found in agricultural experiment stations or plant seed companies for which some history of work or familiarity exists. The populations may be a diverse collection of semi-independent ancestry selected for diverse economic or ecological adaptability, or may be a single land race or traditional variety of proven value. Depending on how narrowly based these populations are with respect to economic or ecological demands, or how little genetic variation may exist in the breeding populations, new populations may often be needed. Ordinarily, populations found in wild forest conditions that are different from the environments of commercial applications require some enhancement before they can be used in advanced breeding programs. This enhancement may take the form of selective breeding within these populations to eliminate deleterious traits or to increase value and adaptability to commercial production. Independent populations are possible to develop as substitutes or competitor varieties. Alternatively, a population may be most easily

enhanced by crossing into established varieties and selecting among later generations of segregants (Frey et al., 1984). Depending on how these enhancement
activities are carried out, genetic diversity within varieties may be increased or
diminished, but so long as populations are kept separate, useful diversity between
varieties may well be increased.

To ensure the continued availability of useful foundation populations, Kannenberg (1984) suggested that hierarchies of populations be established. In these, one
or two base levels of partially enhanced populations feed genes, individuals, or
populations into advanced breeding generations that are more narrowly selected.
Namkoong (1984b) suggested that multiple populations that are simultaneously
developed better serve the needs of breeders of species that have longer breeding
cycles or for which predictability of future needs is poor.

For most forest tree species, there is little distinction between the collection and
evaluation populations and those in which selection enhances the populations
sufficiently for direct use. This often occurs if seed source testing may lead directly
into use of selected trees for commercial seed production. For other species with
possibly more highly developed breeding populations, the stages are more distinct,
and collection may involve only those sources that happened to have been used in
some locality without any purposefully directed sampling of evolutionary diversity.

As long as such collections are secure, either in individual repositories or in
several duplicate sets, there is little need for other repositories. A danger exists,
however, that seed or tissue collections and repositories, or living conservation
collections of plants or animals for these purposes, may destroy existing diversity
by homogenizing all populations. To preserve the sampled diversity, separate
source populations must be kept distinct, or multiple in situ storage areas must be
maintained.

We might therefore characterize the management techniques used for commercially important species as ranging from advanced breeding with its supporting
enhancement populations to targeted collection and evaluation based on ecogeographic surveys. While these may be ex situ programs, in situ methods are often
used for initial surveys, and when areas of use and origin overlap and the ex situ
techniques of seed or tissue collection and storage are either unavailable or insecure,
such as with species with recalcitrant seed, or where ex situ living collections are not
maintainable. In the future, if more species enter commerce, then greater use of in
situ or naturalized ex situ populations will be seen.

The development and use of the primary commercial species are usually based on
substantive biological and economic reasons, and immediate payoffs are to be
expected from investment in these. Nevertheless, as pests and pathogens change, as
climates, soils, and arable land change, and as technologic and economic requirements change, the list of primary varieties and species will ultimately also change.
The rate of change may be on the order of one generation or, it may span geologic
time, depending on the rate of change in ecological or economic forcing factors.
With the quickening pace of global environmental change, what once took eons to
notice may now take decades. The need to develop alternatives originates not only
from shifts in adaptational requirements but also from the need to generate useful
genetic variation in crop species. Populations are therefore needed to serve as

sources for enhanced breeding populations and, either directly or ultimately, for advanced generation breeding.

For these species, the available management techniques are usually less capital intensive, because returns on capital investment are longer in coming and less certain of achievement. Nevertheless, trials of secondary or noncommercial varieties and species often have been established to either replace others entirely or to serve as a source of genes for backcrossing, hybridization, or possibly for gene transfer into established varieties. The uses of in situ and ex situ techniques are different for populations that serve these purposes than for known commercial varieties, and differ among the various applicable management techniques. These programs usually have two principal objectives:

1. Sampling the distribution of genetic variations and understanding present patterns and past cause of that distribution, and
2. Analyzing and discriminating useful differences among genes, individuals, or populations.

The first objective involves targeted sampling and ecogeographic surveys, and is aimed at understanding the distribution of genetic variation for biological insight as well as enabling breeders to sample populations for useful traits or genes more efficiently. As such, it involves studies on the structure of species, the ways in which allelic or gametic frequencies may be heterogeneously distributed, and the extent to which such patterns may have been related to species evolution and may be useful for its future survival or commercial use. One of the problems faced by investigators of population structure is that they can only study existing patterns of variation for at most only a few generations, and these have usually already been altered by destabilizing human and other interventions. If, as earlier suggested, multiple equilibrium states may exist within a complicated global dynamic, then the samples available today do not represent a single stable equilibrium. Thus, the distribution of alleles in their present state does not necessarily represent any natural stable equilibrium or optimal condition. It is more important to track population changes and to allow the species to evolve within an array of environments, primarily in situ, but also possibly in a combination of in situ and ex situ conditions.

The second objective is finding useful variations regardless of mode of origin, and involves ecogeographic surveys with collection and preliminary evaluation in situ. To test wider adaptability, the design and analysis of ex situ tests are well known (e.g., Section 8.2) and, given a sample of populations, the patterns of genetic variation in test environments may be discernible. Such ex situ ecogeographic studies can direct population resampling if the traits desired are not sufficiently well developed in the initial sample, or if it is decided that sufficient gain can be achieved by breeding within previously sampled populations. For some Central American pine species, ex situ stands are the only physically secure means to maintain valuable population sources.

Once endangered sources of germplasm are secured, geneticists trust that testing and enhancement programs can then be carried out. Thus, the source materials for testing and development must be some ex situ or in situ sample of the available gene pool. At a minimum, such a sample must be large enough to have a reasonable chance of saving the useful genes. Even with the best sampling efforts, however, this minimum may miss many alleles if they are rare in the places and time of the sample. To minimize the probabilities of losing alleles, then, sampling must either be sufficiently redundant or directed to a pattern of areas covering populations most likely to yield useful allelic variants. Hence, knowing the population genetic structure of these species helps determine the sampling structure. For species of only potential commercial interest, the sampling will largely be in situ.

To reflect the structural diversity that may exist in populations, targeted samples are needed from different areas, populations, and individuals. It may also be necessary to sample communities in different stages of development, with respect to their own population dynamics or to other nonconstant variables of the environment and mating patterns. However, the only recourse we often have is to sample and maintain sufficiently large populations so that even those alleles that occur at low frequency would have high probability of being saved. Guidelines can be provided for population sizes and distribution, but these must be surrounded by caveats with respect to risks to population integrity and reproduction (see Namkoong, 1985a).

The more information available on the centers of diversity or dispersal patterns, the more efficiently the sampling can be designed. Not only should central populations be sampled, but also populations from extreme sites and isolated locations may contain adaptational features in high frequency. For species in early or secondary succession stages where population structure may easily change over short periods of time, different genotypic distributions may easily exist among different aged stands. A variety of different site-occupancy ages might then also be useful to sample. For species that evolve with competitors, mutualists, or pests and pathogens, a multiplicity of population samples increases the chances of finding populations with higher frequencies of useful but rare alleles. For disease-resistance alleles, it may often be more useful to sample alleles that have coevolved with pathogens, because these may be most useful in breeding for stable host–pathogen systems. There may also exist special indicator variables, physical or biotic, that imply the existence of a substantial frequency of generally rare alleles, and as ecogeographic studies progress, these too may be used to target sampling of in situ populations.

Once the sampling is determined, in situ conservation for collection and evaluation is

1. Necessary when ex situ cannot succeed for biological or technical reasons, such as wild fruit and nut species;
2. Useful when the local environment provides the support factors or desired evolutionary pressure for that species and when population sizes cannot be maintained in ex situ collections, such as the forage species;
3. Desirable when we are ignorant of the species needs.

Ex situ is needed when the natural locations are insecure, or are distant from where the species is needed for development, testing, and enhancement; it is most useful when the natural populations cannot be secured and when ex situ conservation is cheaper and is desirable as a backup for in situ population samples.

Various species require vastly different areas for their conservation, because they are different in individual organism size and dispersal patterns. Different mixtures of in situ and ex situ populations will be needed. It may also sometimes occur that the natural dispersal of a species may be so wide in certain areas that a hundred square kilometers are needed to sample a few hundred adults. In such cases, the conservation of a local population may require more area than can be accommodated. In such cases, transplanting or moving individuals into higher density semimanaged conditions might be considered to condense its distribution within approximately normal environments. This may not be possible for some species to endure, and for many species would entail a selective adaptation to new environmental conditions; however, it may often be useful as a means of capturing the advantages of both in situ and ex situ techniques. This is commonly done with forest trees, and is very useful whenever natural stands are in danger of physical destruction.

In order to ensure the existence of appropriately distributed sample populations, species population sizes and location distributions should be determined and cross-checked with existing reserves and conservation areas. Then, a coordinated plan for supplementary ex situ and in situ conservation areas can be drawn up on the basis of the security of reserved areas for gene conservation purposes.

Several tropical pine species of potential commercial value are included in targeted samples and ecogeographic surveys of in situ stands, but collection and evaluation is heavily in ex situ test plantings. It should be noted that it is not necessary for populations that exist in test or conservation stands to follow all the protocols of commercial breeding. It may often occur that the process is telescoped and test populations immediately become commercial varieties, especially in species without a long history of commercial breeding development. In the future, if the potential commercial value of more species is recognized from present explorations and molecular genetic technologies, in situ sampling may become more important while ex situ testing may be more widely used.

8.4.4 Noncommercial Species

Beyond these few species the vast majority of tree species serve no direct commercial function. However, there are at least three possible functions for these other species which force us to consider their welfare: their future potential for direct use, our need to understand natural population processes, and their support of ecosystems that stabilize other benefits.

The potential for direct use by varietal substitution of traditional species on traditional sites is not high, since foresters have probably effectively screened the available biota. However, for newly understood or developed products or sites, where new uses or processes introduce new utilities, or where new processes change the environment for economic production, previously noncommercial species may become important. With the newer possibilities of genetic engineering, almost any species might contribute useful genes or processes. Because our understanding of the evolution of even the commercial species is so tenuous, it is clearly beneficial to at least maintain a sample of the evolutionary system. By learning about the possible behavior of interacting systems, we can then be informed of the possible ways that forestry can function in a more stable interaction with the biota. The design of future silvicultural systems may lead to more stable ecosystems and to the construction of populations with broader adaptabilities to variable environments.

The contribution of noncommercial species to ecosystem functioning and stability cannot be lightly dismissed. While highly complex and interdependent webs of association may often be fragile and easily degenerated with high extinction rates of component species (May, 1973), the existence of fragility does not imply that species or systems should be allowed to die. Rather, for our own benefit, the various functions of noncommercial species should be understood to include the long-term productivity of all other parts of the ecosystem. Thus, even apparently unimportant species and ecosystems may be critical to the functioning of commercial agriculture, and some most certainly are. Our ignorance of which species and which functions have greater or lesser importance only forces us to consider conservation of more species than are minimally necessary.

The management options for these populations are more restricted than for commercial species and stands. Some form of in situ conservation seems necessary, as for tropical forests, even though it may be neither the most secure nor the least costly option. Because species values are likely to be associated with community functions, conservation is perhaps most easily assured by area management as in reserves, parks, or natural areas. Natural areas are sometimes chosen to maximize species-level diversity, but this is not necessarily sufficient for intraspecies diversity. The requirements of population size and multiple-population dispersal remain the same for each constituent species, and because little direct control of each species in any one area can be expected, sample dispersal and redundancy in size of individual areas and in numbers of separate populations is needed. For species with different population genetic structures, it is not enough to conserve one ecosystem sample. For species that respond to known environmental variables, including the coevolution of other species, sampling from the range of those other variables is necessary for capturing significant genetic variability.

For many species, however, even such recommendations are futile, because so little is known of the species' distribution or even of their existence. For these, targeted sampling in centers of diversity such as the Brazilian Amazon, as outlined by Pires (1978) may be the only realistic hope of their conservation. A minority of the hundreds of species per hectare are known and mapped. Obviously, where known centers of diversity exist within a known species, those would be prime targets for sampling. Even then supplementary sampling from more extreme populations is desirable (Namkoong, 1980a). Similarly, the design of natural reserves should not focus exclusively on centers of origin or diversity, but should also include areas of more extreme habitat of the various biotypes to ensure that the genetic diversity of the contained species is well sampled. Thus, in order to conserve the viability of an ecosystem and to ensure the availability of genetic variations, the dynamics of species evolution generally requires multiple-area and multiple-population sampling.

Whether we consider a species and its associates to have commercial value or not, we know that genetic and ecological variations are not likely to be in an evolutionarily static state. Some may have impoverished gene pools, and some may contain all genetic variations within single large populations, but most must be considered to be in some transient evolutionary state. Whether they have been stable in the recent past or not, human activities have probably changed at least many of their equilibria. For managers of genetic resources, the goal is not to conserve a static state, but to contain a dynamic system, even though our understanding of its dynamics is very meager.

Using in situ reserves seems necessary for the noncommercial species, because of its practicality and the necessity to allow the populations to evolve. This is not to exclude certain ex situ practices such as introducing populations of key species of trees to alter or reestablish an ecosystem. Moreover, the mutual evolution of many different kinds of species at different successional stages and trophic levels requires that conflicting practices often be considered. For pioneer and secondary successional species and the animals dependent on them, some continued disturbance types, sizes, and frequencies are required, whereas climax communities require less disturbance. Because some of these requirements cannot be simultaneously met because of physical or managerial limitations, and for reasons of safety, a multiplicity of sites and management regimes is required. Compartmental management systems can be designed with some compartments kept under strict preservation status, some under frequent disturbance, and other sets with designed juxtapositions of successional stages. Some deliberate intervention will often be necessary to preserve certain processes and species, and targeted sampling of special sites may be needed to supplement any general-purpose network.

Thus, establishing samples of critical ecosystems is but a necessary first step in the establishment of a network of natural and managed areas. While efficiency demands that natural area reserves contain maximum diversity, they alone are not sufficient for conserving much of the biological diversity.

8.4.5 Global Systems

In any global strategy for the conservation of biological diversity for all the values which the multitude of species provides, account must be taken of the mixed objectives for, and the mixed biological status of, the array of species. There is a biological continuum of species, populations, and varieties that fill various commercial and noncommercial needs, and for which various management programs and in situ and ex situ techniques are appropriate. There is also a flow of species and varieties among these categories of use and among management techniques. Thus, there can be no simple distinction made among species or management methods that consistently separates the usefulness of in situ from ex situ techniques. In fact, they often support one another in the same species programs. For example, in situ populations lend security to ex situ collections for breeding, while ex situ seed storage may provide security for in situ conservation populations threatened with local extinction.

Even when not supporting each other, however, different agencies may act on the same species with different objectives and pursue different population support techniques. A global strategy must account for the variety of objectives and management techniques available and consider which problems are most urgent to attack.

References

Adams WT, Roberds JH, Zobel BJ (1973) Intergenotypic interactions among families of loblolly pine (*Pinus taeda* L.). Theor Appl Genet 43:319–322

Ager A, Guries R, Lee C (1982) Genetic gains from red pine seedling seed orchards. In: Proc 28th North Eastern Forest Tree Improvement Conf, Inst Natural Envir Res, July 7–9, 1982, Univ. of New Hampshire, Durham, New Hampshire, USA, 315 pp

Allard RW (1960) Principles of plant breeding. Wiley, New York, 485 pp

Allard RW, Adams J (1969) The role of intergenotypic interactions in plant breeding. Proc XII Int Congr Genet 3:349–370

Anderson TW (1958) An introduction to multivariate statistical analysis. Wiley, New York, 374 pp

Antonovics J (1968) Evolution in closely adjacent plant populations. VI. Manifold effects of gene flow. Heredity 23:507–524

Asmussen MA (1979) Regular and chaotic cycling in models of ecological genetics. Theor Popul Biol 16:172–190

Asmussen MA, Clegg MT (1981) Dynamics of the linkage disequilibrium function under models of gene-frequency hitchhiking. Genetics 99:337–356

Baker RJ (1986) Selection indices in plant breeding. CRC Press, Boca Raton, Florida, USA, 218 pp

Baker RJ (1988) Differential response to environmental stress. In: Proc 2nd Int. Conf Quant Genet, Sinauer Assoc, Sunderland, Massachusetts, USA

Baker LH, Curnow RN (1959) Choice of population size and use of variation between replicate populations in plant breeding programs. Crop Sci 9:555–560

Barnes RD (1986) Multiple population tree breeding in Zimbabwe. In: Proc IUFRO conf. joint meeting of working parties on breeding theory, progeny testing, seed orchards, North Carolina State Univ, Ind Coop Tree Imp Prog, Raleigh, North Carolina, USA, pp 285–297

Bergeman F, Scholz F (1987) The impact of air pollution on the genetic structure of Norway spruce. Silvae Genet 36:80–83

Bishir J, Namkoong G (1987) Unsound seed in conifers: estimation of number of lethal alleles and of the magnitude of external effects. Silvae Genet 36:180–185

Blackith RE, Reyment RA (1971) Multivariate morphometrics. Academic Press, New York, 412 pp

Box GEP, Lucas HL (1959) Design of experiments in non-linear situations. Biometrika 46:77–90

Brotschol JV, Roberds JH, Namkoong G (1986) Allozyme variation among North Carolina populations of *Liriodendron tulipifera* L. Silvae Genet 35:131–138

Burdon, R.D. (1988) Recruitment for breeding populations: objectives, genetics, and implementation. In: Proc 2nd Int Conf Quant Genet, Sinauer Assoc, Sunderland, Massachusetts, USA

Burdon RD, Namkoong G (1983) Multiple populations and sublines. Silvae Genet 32:221–222

Burley J, Burrows PM, Armitage FB, Barnes RD (1966) Progeny test designs for *Pinus patula* in Rhodesia. Silvae Genet 15:166–173

Bush RM, Smouse PE, Ledig FT (1987) The fitness consequences of multiple-locus heterozygosity: The relationship between heterozygosity and growth rate in pitch pine (*Pinus rigida* Mill.). Evolution 41:787–798

Campinhos E, Ikemori YK (1977) Tree improvement program of *Eucalyptus* spp. Preliminary results. In: Proc FAO/IUFRO 3rd World Consult Forest Tree Breeding, CSIRO, Canberra, Australia, pp 717–738

Campbell RK, Echols RM, Stonecypher RW (1985) Genetic variances and interactions in 9-year-old Douglas-fir grown at narrow spacings. Silvae Genet 35:24–32

Chang TT (1985) Principles of genetic conservation. Iowa State J Res 59:325–348

Cheverud J (1982) Phenotypic, genetic, and environmental morphological integration in the cranium. Evolution 36:499–516

Cheverud J (1984) Quantitative genetics and developmental constraints on evolution by selection. J Theor Biol 110:155–171

Cockerham CC (1954) An extension of the concept of partitioning hereditary variance for analysis of covariance among relatives when epistasis is present. Genetics 39:859–882

Cockerham CC, Weir BS (1973) Descent measures for two loci with some applications. Theor Popul Biol 4:300–330

Cockerham CC, Weir BS (1983) Variance of actual inbreeding. Theor Popul Biol 23:85–109

Comstock RE (1977) Quantitative genetics and the design of breeding programs. In: Pollak et al (eds) Proc Int Conf Quant Genetics, August 16–21, 1976, Iowa State Univ Press, Ames, Iowa, USA, pp 705–718

Comstock RE, Robinson HF (1948) The components of genetic variance in populations of biparental progenies and their use in estimating the average degree of dominance. Biometrics 4:254–266

Comstock RE, Robinson HF, Harvey PH (1949) A breeding procedure designed to make maximum use of both general and specific combining ability. Agron J 41:360–367

Conkle, MT (1970) Hybridization—application to pine populations. In: Namkoong and Stern (eds) Second meeting of the working group in quantitative genetics, Southern For Exp Sta, New Orleans, Louisiana, USA, 133 pp

Cox TS, Goodman MM, Murphy JP (in press) The contribution of exotic germplasm to American agriculture. In: Kloppenburger (ed) Seeds and sovereignty. Duke Univ Press, Durham, North Carolina, USA

Crow JF, Kimura M (1970) An introduction to population genetics theory. Harper & Row, New York, 591 pp

Crow JF, Morton NE (1955) Measurement of gene frequency drift in small populations. Evolution 9:202–214

Dawkins R (1982) The extended phenotype. Freeman, Oxford, UK, 307 pp

Delwaulle JC (1985) Plantations clonales d'*Eucalyptus* hybridas an Congo. Rev Bois Forets Trop 208:37–42

Dempster ER (1955) Genetic models in relation to animal breeding. Biometrics 11:535–536

Dorman KW (1976) The genetics and breeding of southern pines. Agric Handbk 471, USDA Forest Serv, Washington, DC, 407 pp

Dudley JW (1977) 76 generations of selection for oil and protein percentage in maize. In: Pollak, Kempthorne, and Bailey Jr (eds) Proc Int Conf Quant Genetics, August 16–21, 1976, Iowa State Univ Press, Ames, Iowa, USA, pp 459–473

Eberhart SA, Russell WA (1966) Stability parameters for comparing varieties. Crop Sci 6:36–40

Emigh TH, Pollak E (1979) Fixation probabilities and effective population numbers in diploid populations with overlapping generations. Theor Popul Biol 15:86–107

Eriksson G, Gullberg U, Kang H (1984) Breeding strategy for short rotation woody species. In: Perttu (ed) Ecology and management of forest biomass production systems. Dept Ecol Environ Res, Swed Univ Agric, Sci Rep 15:199–216

Falconer DS (1981) Introduction to quantitative genetics. Longman, London, 340 pp

Finlay KW, Wilkinson GN (1963) The analysis of adaptation in a plant-breeding programme. Aust J Agric Res 14:742–754

Fisher R (1949) The theory of inbreeding. Academic Press, New York, 150 pp

Fowler DP (1978) Population improvement and hybridization. Unasylva 30:21–26

Fowler DP, Morris RW (1977) Genetic diversity in red pine: evidence for low genic heterozygosity. Can J For Res 7:343–347

Frankham R, Jones LP, Barker JSF (1968a) The effects of population size and selection intensity in selection for a quantitative character in Drosophila. I. Short-term response to selection. Genet Res 12:237–248

Frankham R, Jones LP, Barker JSF (1968b) The effects of population size and selection intensity in selection for a quantitative character in Drosophila. II. Long term response to selection. Genet Res 12:249–266

Frankham R, Jones LP, Barker JSF (1968c) The effects of population size and selection intensity in selection for a quantitative character in Drosophila. III. Analyses of the lines. Genet Res 12:267–283

Franklin EC (1968) Artificial self-pollination and natural inbreeding in Pinus taeda L. Diss Abstr B Sci Eng 29(4):1225

Franklin EC (1979) Model relating levels of genetic variance to stand development of four North American conifers. Silvae Genet 28:207–212

Franklin I, Lewontin RC (1970) Is the gene the unit of selection? Genetics 65:707–734

Freeman GH, Perkins JM (1971) Environmental and genotype-environmental components of variability. VIII. Relations between genotypes grown in different environments and measures of these environments. Heredity 27:15–23

Frey KJ, Cox TS, Rodgers DM, Bramel-Cox P (1984) Increasing cereal yields with genes from wild and weedy species. In: Proc 15th Int Congr Genetics, Vol 4, New Delhi, India, pp 51–68

Friedman, ST (1987) Natural regeneration: Genetic and seed aspects. USDA Forest Service, Pacific Northwest Region, Portland, Oregon, USA, 20 pp

Friedman ST (1981) Estimation of gene flow into the seed orchards of loblolly pine. Theor Appl Genet 69:609–615

Friedman ST, Namkoong G (1986) Estimating family means using unbalanced incomplete blocks. In: Proc IUFRO Conf joint meeting of working parties on breeding theory, progeny testing, seed orchards. North Carolina State Univ, Ind Coop Tree Imp Prog, Raleigh, North Carolina, USA, pp 457–468

Geburek Th, Scholz F, Knabe W, Vornweg A (1987) Genetic studies by isozyme gene loci on tolerance and sensitivity in an air pollution Pinus sylvestris field trial. Silvae Genet 36:49–53

Gibson GL (1982) Genotype–environment interaction in Pinus caribaea. Commonwealth For Inst, Oxford, England, 112 pp

Giesel JT (1971) The relations between population structure and rate of inbreeding. Evolution 25:491–496

Gill JL (1965) Selection and linkage in simulated genetic populations. Aust J Biol Sci 18:1171–1187

Ginzburg LR (1983) Theory of natural selection and population growth. Benjamin Cummings, Menlo Park, California, USA, 160 pp

Glock H, Gregorius H-R (1986) Genotype–environment interaction in tissue cultures of birch. Theor Appl Genet 72:477–482

Goodman MM (1984) An evaluation and critique of cereal germ plasm programs. In: Conservation and utility of exotic germmplasm to improve varieties. Pioneer Hi-Breed Int, Des Moines, Iowa, USA, pp 197–251

Gregorius H-R (1984) Functional fitness in exclusively sexually reproducing populations. J Theor Biol 111:205–229

Gregorius H-R, Namkoong G (1983) Conditions for protected polymorphism in subdivided plant populations: 1. uniform pollen dispersal. Theor Popul Biol 24:252–267

Gregorius H-R, Namkoong G (1986) Joint analysis of genotypic and environmental effects. Theor Appl Genet 72:413–422

Gregorius H-R, Namkoong G (1987) Resolving the dilemmas of interaction, separability, and additivity. Math Biosci 85:51–69

Griffing B (1960) Theoretical consequences of truncation selection based on the individual phenotype. Aust J Biol Sci 13:307–343

Griffing B (1967) Selection in reference to biological groups. I. Individual and group selection applied to populations of unordered groups. Aust J Biol Sci 20:127–139

Gullberg U, Kang H (1985a) A model for tree breeding. Stud For Suec 169:1–8

Gullberg U, Kang H (1985b) Application of model for tree breeding to conifers in southern Sweden. Stud For Suec 170:1–8

Hadley G (1964) Nonlinear and dynamic programming. Addison-Wesley, Reading, Massachusetts, USA, 484 pp

Hall RB, Miksche JD, Hansen KM (1976) Nucleic acid extraction, purification, reannealing, and hybridization methods. In: Miksche (ed) Modern methods in forest genetics. Springer-Verlag, Berlin, pp 19–48

Hallauer AR, Miranda JB (1981) Quantitative genetics in maize breeding. Iowa State Univ Press, Ames, Iowa, USA 486 pp

Hamrick JL (1983) The distribution of genetic variation within and among natural plant populations. In: Schonewald-Cox, Chambers, MacBryde, and Thomas (eds) Genetics and conservation. Benjamin Cummings, Menlo Park, California, USA pp 335–347

Hamrick JL, Schnabel A (1985) Understanding the genetic structure of plant populations: some old problems and a new approach. In: Gregorius (ed) Population genetics in forestry, Lecture Notes in Biomathematics, Vol 60. Springer-Verlag, Berlin, pp 50–70

Hanson WD (1970) Genotypic stability. Theor Appl Genet 40:226–231

Hedrick PW (1980) Hitchiking: A comparison of linkage and partial selfing. Genetics 94:791–808

Hedrick PW (1983) Genetics of population. Van Nostrand Reinhold, New York, 629 pp

Hedrick PW (1987) Genetic polymorphism in heterogeneous environments: A decade later. Annu Rev Ecol Syst 17:535–566

Heybroek HM, Stephan BR, von Weissenberg K (1981) Proc Third Int Workshop on Genetics of Host–Parasite Interactions in Forestry, Wageningen, Netherlands, September 14–21, 1980

Hill WG (1969) On the theory of artificial selection in finite population. Genet Res (Cambridge) 13:143–163

Hill WG (1972) Estimation of realised heritabilities from selection experiments. I. Divergent selection. Biometrics 28:747–765

Hill WG (1972) Effective size of populations with overlapping generations. Theor Popul Biol 3:278–392

Hill WG (1982) Predictions of response to artificial selection from new mutations. Genet Res 40:255–278

Hill WG, Rasbash J (1986a) Models of long term selection in finite populations. Genet Res (Cambridge) 48:41–50

Hill WG, Rasbash J (1986b) Models of long-term artificial selection in finite population with recurrent mutation. Genet Res (Cambridge) 48:125–131

Hühn M (1969) Untersuchungen zur Konkurrenz zwischen vershiedenene Genotypen in Pflanzenbestanden. I. Modifikation der Methode von Sakai zur Schatzung der genetischen-, Umwelt- und Konkurrenzvarianz einer Population. Silvae Genet 18:186–192

Hühn M (1970) The competitive environment and its genetic reaction variations. In: Second meeting of the working group on quantitative genetics, Section 22, IUFRO, 1969. USDA For Serv, Southern For Exp Sta, New Orleans, Louisana, USA, pp 62–86

Hühn M (1985) Theoretical studies on the necessary number of components in mixtures. 1. Number of components and yield stability. Theor Appl Genet 70:383–389

Hühn M (1986a) Theoretical studies on the necessary number of components in mixtures. 2. Number of components and yielding-ability. Theor Appl Genet 71:622–630

Hühn M (1986b) Theoretical studies on the necessary number of components in mixtures. 3. Number of components and risk considerations.Theor Appl Genet 72:211–218

Hyun S (1974) The expression of heterosis of improved hybrid poplars in Korea being influenced by the site and the cultural method. In: Proc IUFRO joint meeting of working parties on population genetics, breeding theory, and progeny testing, 30 pp

Jain SK, Bradshaw AD (1966) Evolutionary divergence among adjacent plant populations. Heredity 21:407–441

Johnson RA, Wichern DW (1982) Applied multivariate statistical analysis. Prentice-Hall, Englewood Cliffs, New Jersey, 594 pp

Jones LP, Frankham R, Barker JSF (1968) The effects of population size and selection intensity in selection for a quantitative character in Drosophila. I. Long-term response to selection. Genet Res 12:249–266

Kang H (1980) Designing a tree breeding system. In: Proc 17th Meet Canadian Tree Improvement Assoc, Gander, Newfoundland, Canada, August 27–30, 1979, pp 51–63

Kang H (1982) Components of a tree breeding plan. In: Proc IUFRO joint meeting of working parties on genetics about breeding strategies including multiclonal varieties, Sensenstein, FRG, September 6–10, 1982, pp 119–135

Kang H (1983) Limits of artificial selection under balanced mating systems with family selection. Silvae Genet 32:188–195

Kang H, Namkoong G (1979) Limits of artificial selection under balanced mating systems. Silvae Genet 28:53–60

Kang H, Namkoong G (1980) Limits of artificial selection under unbalanced mating systems. Theor Appl Genet 58:181–191

Kang H, Namkoong G (1988) Inbreeding effective population size under some artificial selection schemes. I. Linear distribution of breeding values. Theor Appl Genet 75:333–339.

Kang H, Nienstaedt H (1987) Managing long-term tree breeding stock. Silvae Genet 36:30–39

Kannenberg LW (1984) Utilization of genetic diversity in crop breeding. Plant genetic resources: a conservation imperative. In: Yeatman, Kafton and Wilkes (eds) AAAS Selected Symp 87. Westview Press. Boulder, Colorado, USA, pp 93–110

Karlin S, Feldman MW (1970) Linkage and selection: two locus symmetric viability model. Theor Popul Biol 1:39–71

Kellison RC (1966) A geographic variation study of yellow-poplar (*Liriodendron tulipifera* L.) within North Carolina. MS Thesis, North Carolina State Univ, Raleigh, North Carolina, USA, 70 pp

Kellison RC (1970) Phenotypic and genotypic variation of yellow poplar (*Liriodendron tulipifera* L.). PhD Thesis, North Carolina State Univ, Raleigh, North Carolina, USA, 112 pp

Kendall MG (1961) A course in multivariate analysis. Hafner, New York, 185 pp

Kendall MG, Stuart A (1963) The advanced theory of statistics, Vol I. Hafner, New York, 433 pp

Kendall MG, Stuart A (1966) The advanced theory of statistics, Vol III. Hafner, New York, 552 pp

Kimura M (1962) On the probability of fixation of mutant genes in a population. Genetics 47:713–719

Kimura M (1983) The neutral theory of molecular evolution. Cambridge Univ Press, Cambridge, England, 367 pp

Kimura M, Crow JF (1963) The measurement of effective population number. Evolution 17:279–288

King JP (1965) Seed source x environment interactions in Scotch pine. I. Height growth. Silvae Genet 14:105–115

Knight R (1970) The measurement and interpretation of genotype–environment interactions. Euphytica 19:225–235

Kojima K, Kellherer T (1963) Selection studies of quantitative traits with laboratory animals. In: Statistical genetics and plant breeding. Natl Acad Sci, Natl Res Counc Publ 982, Washington, DC, pp 395–422

Lande R (1980) The genetic covariance between characters maintained by pleiotropic mutations. Genetics 94:203–215.

Lande R (1984) The genetic correlation between characters maintained by selection, linkage and inbreeding. Genet Res (Cambridge) 44:309–320

Langlet O (1963) Patterns and terms of intra-specific ecological variability. Nature (Lond) 200:347–348

Larsen CS (1951) Genetics in silviculture (trans Anderson ML). Oliver and Boyd, Edinburgh, Scotland, 224 pp

Ledig FT (1970) Genotype x environment interaction in controlled environments: the physiological basis for differential response. In: Second meeting of the working group on quantitative genetics, section 22, IUFRO 1969. USDA For Serv, Southern For Exp Sta, New Orleans, Louisian, USA, pp 90–99

Ledig FT, Fryer JH (1972) A pocket of variability in *Pinus rigida*. Evolution 26:259–266

Ledig FT, Guries RP, Bonefeld BA (1983) The relation of growth to heterozygosity in pitch pine. Evolution 37:1227–1238

Ledig FT, Perry TO (1967) Variation in photosynthesis and respiration among loblolly pine progenies. In: Proc Ninth Southern Conf Forest Tree Improvement, pp 120–128

Lerner IM (1958) The genetic basis of selection. Wiley, New York, 298 pp

Levin SA, Udovic JD (1977) A mathematical model of coevolving populations. Am Nat 111:657–675

Lewontin RC (1974) Annotation: the analysis of variance and the analysis of causes. Am J Hum Genet 26:400–411

Lewontin RC, Kojima K (1960) The evolutionary dynamics of complex polymorphisms. Evolution 14:458–472

Lindgren D, Gregorius H-R (1976) Inbreeding and coancestry. In: Proc Joint Meet Advanced Generation Breeding, Bordeaux, France, pp 49–55

Lowe WJ, van Buijtenen JP (1981) Tree improvement philosophy and strategy for the western Gulf Tree Improvement Program. In: Proc 15th North Am Quant For Gen Workshop, pp 43–50

Matzinger DF, Wernsman EA (1968) Four cycles of mass selection in a synthetic variety of an autogamous species *Nicotiana tabacum*. Crop Sci 8:239–243

May RM (1973) Stability and complexity in model ecosystems. Monographs in population biology, No 6. Princeton Univ Press, Princeton, New Jersey, USA

Mayo O (1980) The theory of plant breeding. Clarendon Press, Oxford, England 293 pp

McCutchan BG, Ou JX, Namkoong G (1985) A comparison of planned unbalanced designs for estimating heritability in perennial tree crops. Theor Appl Genet 71:536–544

Mendenhall W, Schaeffer RL, Wackerly DD (1981) Mathematical statistics with applications. Duxbury Press, Boston, Massachusetts, 686 pp

Mitchell-Olds T, Rutledge JJ (1986) Quantitative genetics in natural plant populations: a review of the theory. Am Nat 127:379–402

Mitton JB (1978) Relationship between heterozygosity for enzyme loci and variation of morphological characters in natural populations. Nature (Lond) 273:661–662

Moll RH, Hanson WD (1984) Comparison of effects of intrapopulation vs interpopulation selection in maize. Crop Sci 24:1047–1052

Moll RH Robinson HF (1967) Quantitative genetic investigations of yield of maize. Zuchter 37:191–199

Moll RH, Stuber CW (1971) Comparisons of responses to alternative selection procedures initiated with two populations of maize (*Zea mays* L.). Crop Sci 11:706–711

Moll RH, Cockerham CC, Stuber CW, Williams WP (1978) Selection responses, genetic-environmental interactions, and heterosis with recurrent selection for yield in maize. Crop Sci 18:641–645

Morrison DF (1976) Multivariate statistical methods, 2d ed. McGraw-Hill, New York, 415 pp

Müller-Stark G (1985) Reproduction success of genotypes of *Pinus sylvestris* L. in different environments. In: Gregorius (ed) Population genetics in forestry. Lecture Notes in Biomathematics, Vol 60. Springer-Verlag, Berlin, pp 118–137.

Murphy J, Kang H (1985) The role of genetic models for silvicultural systems. Proc 1985, Natl Silviculture Workshop, success in silviculture, May 13–17, Rapid City, South Dakota, USA, pp 289–295

Namkoong G (1965) Inbreeding effects on estimation of genetic additive variance. For Sci 12:8–13

Namkoong G (1969) Non-optimality of local races. In: Proc 10th Southern Conf Forest Tree Breeding. Texas For Serv, Texas A&M Univ Press, College Station, Texas, USA, pp 194–197

Namkoong G (1970) Optimum allocation of selection intensity in two stages of truncation selection. Biometrics 26:465–476

Namkoong G (1976) Multiple-index selection strategy. Silvae Genet 25:199–201

Namkoong G (1979) Introduction to quantitative genetics in forestry. US Dept Agric Tech Bull 1588, 342 pp

Namkoong G (1980a) Genetic considerations in management of rare and local tree species. Proc conf dendrology in the eastern deciduous forest biome, Blacksburg, Virginia, USA, pp 59–66

Namkoong G (1980b) Breeding for variable environments. For Ind Lect Ser 6, Forestry Program, Univ Alberta, Edmonton, Alberta, Canada

Namkoong G (1984a) Genetic structure of forest tree populations. Proc 15th Int Cong Genetics, December 12–21, 1983, New Delhi, India, pp 351–360

Namkoong G (1984b) Strategies for gene conservation in forest tree breeding. In: Yeatman, Kafton and Wilkes (eds) Plant gene resources: a conservation imperative. AAAS Selected Symp 87, Westview Press, Boulder, Colorado, USA, pp 79–92

Namkoong G (1985a) The population genetic basis of breeding theory. In: Gregorius (ed) Population genetics in forestry. Lecture Notes in Biomathematics, Vol. 60, Springer-Verlag, New York, pp 2–15

Namkoong G (1985b) The influence of composite traits on genotype by environment relations. Theor Appl Genet 70:315–317

Namkoong G, Bishir J (1987) The frequency of lethal alleles in forest tree populations. Evolution 41:1123–1127

Namkoong G, Conkle MT (1976) Time trends in genetic control of height growth in ponderosa pine. For Sci 22:2–12

Namkoong G, Gregorius H-R (1985) Conditions for protected polymorphisms in subdivided plant populations. 2. Seed versus pollen migration. Am Nat 125:521–534

Namkoong G, Johnson JA (1987) Influence of the value function on genotype-by-environment relations. Silvae Genet 36:92–94

Namkoong G, Roberds J (1974) Choosing mating designs to efficiently estimate genetic variance components for trees. I. Sampling errors of standard analysis of variance estimators. Silvae Genet 23:43–53

Namkoong G, Roberds JH (1982) Short-term loss of neutral alleles in small population breeding. Silvae Genet 31:1–6

Namkoong G, Selgrade JF (1986) Frequency-dependent selection in logistic growth models. Theor Popul Biol 29:64–86

Namkoong G, Barnes RD, Burley J (1980) A philosophy of breeding strategy for tropical forest trees. Trop For Paper No. 16, Commonwealth For Inst, Univ Oxford, England, 67 pp

Namkoong G, Snyder EB, Stonecypher RW (1966) Heritability and gain concepts for evaluating breeding systems such as seedling orchards. Silvae Genet 15:76–84

Namkoong G, Usanis RH, Silen RR (1972) Age-related variation in genetic control of height growth in Douglas-fir. Theor Appl Genet 42:151–159

Nance WL, Wells O (1981) Site index models for height growth of planted loblolly pine (*Pinus taeda* L.) seed sources. In: Proc 16th Southern Forest Tree Improvement Conf, May 27–28, 1981, Blacksburg, Virginia, USA, pp 88–96

Nanson A (1970) Juvenile and correlated trait selection and its effect on selection programs. Second meeting of the working group on quantitative genetics, section 22, IUFRO 1969. USDA For Serv, Southern For Exp Sta, New Orleans, Louisiana, USA, pp 17–26

Nienstaedt H, Kang H (1983) A budget tree improvement program: an example. USDA For Serv, North Central For Exp Sta, Res Note NC-294

O'Malley DM, Bawa K (1987) Mating system of a tropical rain forest tree species. Am J Bot 74:1143–1149

Orr-Ewing AL (1965) Inbreeding and single crossing in Douglas-fir. For Sci 11:279–290

Owen G (1968) Game theory. WB Saunders, Philadelphia, Pennsylvania, USA 228 pp

Paques LE (1984) Tree improvement strategies: modelization and optimization. MSc Thesis, North Carolina State Univ, Raleigh, North Carolina, USA, 99 pp

Park YS, Fowler DP (1988) Genetic variances among clonally propagated populations of tamarack and the implications for clonal forestry. Can J For Res (in press)

Patel RM Cockerham CC, Rawlings JO (1962) Selection among factorially classified variables. North Carolina State Univ, Inst Stat Mimeo Ser 317, Raleigh, North Carolina, USA, 43 pp

Patel RM, Cockerham CC, Rawlings JO (1969) Selection among factorially classified variables. Biometrics 25:49–61

Pepper WD (1983) Choosing plant-mating design allocations to estimate genetic variance components in the absence of prior knowledge of the relative magnitudes. Biometrics 39:511–521

Perkins JM, Jinks JL (1968) Environmental and genotype-environmental components of variability. III. Multiple lines and crosses. Heredity 23:339–356

Pires JM (1978) The forest ecosystems of the Brazilian Amazon. Tropical forest ecosystems. UNESCO-UNEP, Vendome, France, pp 607–627

Pooni HS, Jinks JL (1980) Non-linear genotype × environment interactions. II. Statistical models and genetical control. Heredity 45:389–400

Rauter RM (1982) Recent advances in vegetative propagation including biological and economic considerations and future potential. In: Proc IUFRO joint meeting of working parties in genetics and breeding strategies including multiclonal varieties. Lower Saxony For Res Inst, Escherode, Germany, pp 33–58

Rawlings JO (1970) Present status of research on long and short-term recurrent selection in finite populations—choice of population size. In: Second meeting of the working group on quantitative genetics, section 22, IUFRO 1969. USDA For Serv, Southern For Exp Sta, New Orleans, Louisiana, USA, pp 1–15

Rehfeldt J (1984) Microevolution of conifers in the northern Rocky Mountains. A view from common gardens. In: Proc 8th North Am For Biol Workshop, Dept For Res, Utah State Univ, Logan, Utah, USA, pp 132–146.

Roberds JH, Conkle MT (1984) Genetic structure in loblolly pine stands: allozyme variation in parents and progeny. For Sci 30:319–329

Roberds JH, Namkoong G (1986) Maximization of expected value for a trait in an environmental gradient. In: Tauer and Hennessey (eds) Physiologic and genetic basis of forest decline, Proc Ninth North Am For Biol Workshop, June 15–18, 1986, Oklahoma State University, Stillwater, Oklahoma, USA. Oklahoma State Univ Publ #AI-6154-0686, RO

Roberds JH, Hain FP, Nunnally LB (1987) Genetic structure of southern pine beetle populations. For Sci 33:52–69

Roberds JH, Namkoong G, Davey CB (1976) Family variation in growth response of loblolly pine to fertilizing with urea. For Sci 22:291–299

Robertson A (1960) A theory of limits in artificial selection. Proc R Soc Lond B Biol Sci 153:234–249

Robertson A (1961) Inbreeding in artificial selection programmes. Genet Res (Cambridge) 2:189–194

Roughgarden J (1979) Theory of population genetics and evolutionary ecology: An introduction, MacMillan, New York, 634 pp

Rudolph TD (1981) Four-year height growth variation among and within SO, S1 × S1, S1 open-pollinated, and S2 inbred jack pine families. Can J For Res 11:654–661

Rudolph TD, Yeatman CW (1982) Genetics of jack pine. USDA For Serv Res Pap WO-38, 60 pp

Rudolph TD, Kang H, Guries R. Selection for juvenile height and early flowering in jack pine. I. Realized genetic gain in juvenile height (submitted for publication)

Sakai K (1955) Competition in plants and its relation to selection. Cold Spring Harbor Symp Quant Biol 20:137–157

Sakai K (1965) Contributions to the problem of species colonization from the viewpoint of competition and migration. In: Baker and Stebbins (eds) Genetics of colonizing species. Academic Press, New York, pp 215–241

Sakai K, Mukaide H (1967) Estimation of genetic, environmental, and competitional variances in standing forests. Silvae Genet 16:149–152

Sakai K, Mukaide H, Tomita K (1968) Intraspecific competition in forest trees. Silvae Genet 17:1–5

Schutz WM, Brim CA (1971) Inter-genotypic competition in soybeans. III. An evaluation of stability in multiline mixtures. Crop Sci 11:684–689

Schutz WM, Usanis SA (1969) Intergenotypic competition in plant populations. II. Maintenance of allelic polymorphisms with frequency dependent selection and mixed selfing and random mating. Genetics 61:875–891

Selgrade JF, Namkoong G (1984) Dynamical behavior of differential equation models of frequency and density dependent populations. J Math Biol 19:133–146

Snyder EB (1966) Lattice and compact family block designs in forest genetics. In: Joint proceedings second genetics workshop of the Society of American Foresters and the seventh improvement conference 1965. US For Serv Res Pap NC-6. North Central For Exp Sta, St. Paul, Minnesota, USA, pp 12–17

Snyder EB (1969) Parental selection versus half-sib family selection of longleaf pine. Proc 10th Southern For Tree Imp Conf, Houston, Texas, USA National Technical Inf. Serv. Springfield, Virginia, USA, pp 84–88

Snyder EB, Allen RM (1971) Competitive ability of slash pine analyzed by genotype × environment stability method. Proc 11th Southern For Tree Imp Conf, 1971, Georgia. Atlanta, Nat'l. Tech. Inf. Serv. Springfield, Virginia, USA, pp 142–147

Squillace AE (1970) Genotype–environment interactions in forest trees. In: Second meeting of the working group on quantitative genetics, section 22, IUFRO 1969. USDA For Serv, Southern For Exp Sta, New Orleans, Louisiana, pp. 49–61

Steel RGD, Torrie JH (1980) Principles and procedures of statistics: a biometrical approach. McGraw-Hill, New York, 633 pp

Stern K, Roche L (1974) Genetics of forest ecosystems. Springer-Verlag, Berlin, 330 pp

Strauss SH, Libby WJ (1987) Allozyme heterosis in radiata pine is poorly explained by overdominance. Am Nat 130:879–890

Tai GCC (1971) Genotypic stability analysis and its application to potato regional trials. Crop Sci 11:184–190

Thompson G (1977) The effects of a selected locus on linked neutral loci. Genetics 85:753–788

Tigerstedt PMA (1984) Genetic mechanisms for adaptation: the mating system of Scots pine. Proc 15th Int Congr Genetics, December 12–21, 1983, New Delhi, India, pp 317–322

Toda R (1956) On the crown slenderness in clones and seedlings. Z Forstgenet Forstpflanzenaucht 5:1–5

Toda R (ed) (1974) Forest tree breeding in the world. Govt For Exp Sta Japan, Meguro, Tokyo, 205 p

Ununger J (1987) Dynamics of juvenile growth characters in Norway spruce (*Picea abies* (L.) Karst) full-sib families. Res Notes 39, Swedish Univ Agric Sci, Dept For Genetics, Uppsala, Sweden, 85 pp

van Buijtenen JP, Lowe WJ (1979) The use of breeding groups in advanced generation breeding. Proc 15th Southern For Tree Imp Conf, Gulfport, Mississippi, USA, pp 59–65

Weir BJ, Cockerham CC (1974) Behavior of pairs of loci in finite monoecious populations. Theor Popul Biol 6:323–354

Wells OO, Wakeley PC (1966) Geographic variation in survival, growth, and fusiform-rust infection of planted loblolly pine. Soc Am For, For Sci Monogr 11, 40 pp

Williams JS (1962) The evaluation of a selection index. Biometrics 18:375–393

Wisconsin DNR (1987) Nursery tree distribution and tree planting report 1987. Wisconsin Dep Nat Resour, Madison, Wisconsin, USA, 9 pp

Wright JW (1976) Introduction to forest genetics. Academic Press, New York, 463 pp

Wright S (1922) Coefficients of inbreeding and relationship. Am Nat 56:330–338

Wright S (1931) Evolution in Mendelian populations. Genetics 16:97–159

Yeh FC (1981) Analyses of gene diversity in some species of conifers. In: Conkle (ed) Proc Symp on Isozymes of North American Forest Trees and Forest Insects. Pacific Southwest For and Range Exp Sta, For Serv, USDA, Berkeley, California, USA, pp. 48–52

Yeh FC, Heamon JC (in press) Blocked disconnected diallels: estimating genetic parameters of height growth in 7-year-old coastal Douglas-fir. Can J For Res

Zobel BJ, Talbert J (1984) Applied forest tree improvement. Wiley, New York, 505 pp
Zsuffa L (1979) The features and prospectus of poplar breeding in Ontario. In: Coyle, Zsuffa and Anderson (eds) Poplar research, management and utilization in Canada. For Res Inform Paper No 102, Ministry of Natural Resources, Ontario, Canada, pp 1–6

Index

Adaptability, 47, 51
Allele, 16, 49, 156
    allelic effect, 13, 14, 22, 117
    deleterious, 41, 57, 58
    frequency, 15, 17, 18, 86, 90
    loss, 27, 32, 41, 49, 124, 156
Amazon, 158
Analysis of Covariance (ANCOVA), 80, 81
Analysis of Variance (ANOVA), 80, 81,
    105–107

Backcrossing, 128, 153
Best Linear Unbiased Predictor (BLUP), 76
Breeding
    cycle, 42
    hybrids, 56
    long-term, 38–41, 46, 59, 61
    multiple-generation, see Multiple-genera-
        tion breeding
    multiple-population, 56, 70–74, 97, 98
    programs, 20, 41, 44, 46, 51, 52, 133
    short-term, 40, 41, 54, 59, 61, 83–85
    single-population, 57–66
    techniques, 38, 98–102
    theory, 6
    value, 44, 45, 53, 56, 64
    zones, 52, 53
British Columbia Forest Service Douglas-fir
    program, 72

Canonical Correlation Analysis, 137
Chloroplasts, 11
Chromosome, 86
Cline, 5, 105
Clonal propagation, 68, 70, 77, 79
Clonal propagule, 41, 45, 53
Coevolution, 156
Collinearity, 137
Combining ability
    general (GCA), 44, 66, 102, 148

hybrid, 3
    specific (SCA), 44, 64, 66, 68, 102, 122,
        126
Competition, 115–119
Conservation, 2, 6, 132–159
Controlled crosses, see Pollination, con-
        trolled
Correlation, 79–89; see also Juvenile–mature
        correlations
    coefficient, 85
    genetic, 80, 81, 84, 86–89
    matrix, 137
    phenotypic, 79, 80
Covariance, 75, 76, 79, 80–84, 86, 88, 117,
        136
    of dominance terms, 82
    of epistatic terms, 82
    genetic, 79, 81, 82
Cryptomeria, 118

Deme
    mating, 19, 31, 32, 149
    selection, 19, 149
Deoxyribonucleic acid, see DNA
Determined stem growth, 85
Diallel, 63
    full, 60
    partial (PD), 61, 65, 66, 102
    partial disconnected, 65, 66
    partial factorial, 65, 66
Dioecious, 32, 79
Disequilibria, 16
DNA (deoxyribonucleic acid), 4, 11, 24,
        124
Domestication, 132, 133
Dominance, 13, 148
    complete, 31
    partial, 31, 58
Douglas-fir, 105, 132
Drift, 32, 35
Drosophila, 24

Ecogeographic survey, 153–157
Effects, *see* Gene effects
Environment
  effects, 76, 80, 105, 108–110
  factors, 75
  multiple, 125–127
  variation, 22, 23, 34, 45, 51, 75, 76, 112
Enzyme, 11
Epistasis, 15, 20, 21, 24, 30, 49, 56
Equilibria, 16, 21
Espacement, 115
Evolution, 39, 151
Experimental design, 76, 81, 82, 107, 144
Extinction, 27

Family
  effects, 105
  means, 76, 101
  variance, 78
F coefficient, *see* Inbreeding, coefficient
Fertility, 34
Fisher's Fundamental Theorem of Natural
    Selection, 18, 19, 75
Fitness, 3, 17–20, 57
Fitness values, 30
Fixation, 27, 30, 35, 49, 61, 64

Gain, 5, 42, 54, 62
  achieved, 28
  economic, 38–40, 92, 133
  expected, 46, 48, 62, 64, 77, 78, 93, 95
  generational, 28, 46
  genetic, 40, 93
  long-term, 6, 30, 67, 133
  maximizing, 78, 92, 96, 133
  phenotypic, 28
  predicted, 28, 46, 75
  short-term, 6, 30, 38–40, 133
Gene
  action, 2, 6, 11–16, 31, 56, 98–102
    additive, 57, 67, 74, 86, 87, 114
    dominance, 67, 99, 126
    epistatic, *see* Epistasis
    models, 6, 12–17; *see also* One-locus
      model, Multiple-locus model
    nonadditive, 56, 74, 98, 121
    overdominance, 122, 126
  effects, 7, 11–36, 45, 75, 79
    additive, 26, 66, 106, 148

  fixed, 7
  multiple, 4
  nonadditive, 64, 106, 148
  expression, 11
  frequency, 13–15, 17, 20
  insertions, 11
  products, 11, 49
  structure, 11
  substitution, 11, 12
  transfer, 124
Genetic engineering, 157
Genetic variation, 14–19, 31, 46, 54, 75,
    84, 101; *see also* Variation
  additive, 13–15, 18, 22, 42, 45, 54, 77,
    82, 126
  dominance, 13–15, 22, 126
  estimation of, 22–24, 107
  interfamily, 54
  of jack pine, 38
  levels of, 4, 9
  of natural populations, 143, 150, 155
  nonadditive, 77
  of red pine, 38
  of selected populations, 29
  in species, 2, 3
Genome, 11, 12
Genotype, 53, 54, 107
Genotype by environment interaction, 82,
    105–107, 125
Genotypic value, 23
Griffing's notation, 27
Growth rhythms, 85

Haploid doubling, 58
Hardy–Weinberg frequencies, 14–17, 34
Harmonic mean, 34
Heritability (*h*), 29, 30, 48, 51, 53, 62, 74–
    77, 84, 125
  broad-sense, 23, 54
  change in, 47, 92
  degree of, 6, 51
  estimated, 29, 75–77
  low, 122, 123
  narrow-sense, 24, 54
  realized, 28, 29
Heterosis, *see* Overdominance
Heterozygosity, 4, 58, 122
Heterozygote, 21, 25, 31
Hitchhiking, 89
Homozygosity, 26, 57

Homozygote, 21
Hybrid, 3, 16, 68
  breeding, 66–70, 121, 125, 126
  interspecific hybrids, 2, 122
  progeny, 44
  vigor, *see* Overdominance

Inbred lines, 38, 57, 58
Inbreeding, 15, 31, 32, 63, 89, 122
  coefficient, 32, 33, 34, 60, 61
  depression, 31, 41, 57–59
  effective number, 33, 60, 61
Independence of gene action, 14, 31, 88
Independent effects, 20, 21, 24, 28
Insertions, gene, *see* Gene insertions
Interaction, 109, 117, 144, 146; *see also*
      Genotype by environment interaction
Interlocus effects, 21
Intralocus effects, 27
Isoenzymes, 4, 133, 149

Juvenile
  expression, 83
  –mature correlations, 83, 85
  performance, 79, 84

Linear regression, 106
Linkage, 24, 28, 30, 31, 89
  disequilibrium, 15, 24, 31, 32, 49, 85–87
Locus, 86, 89

Maize, 24, 47, 48, 53, 57, 58, 68, 69, 85
Mating, 41
  designs, 57–66; *see also* Diallel, Tester
      parent
    circular, 62, 64
    cousin, 62, 63, 65
    factorial, 61, 63, 65, 102
    single-pair mating (PM), 61, 63
Mean
  arithmetic, 34
  geometric, 82
  harmonic, 34
Migration, 36, 149
  rates, 35
Mitochondria, 11
Monoculture, 55

Monoecious, 32, 42, 59
Mother trees, 2
Multiple-generation breeding, 38, 39, 54
Multiple-locus model, 14, 15, 24, 27
Multiple-species strategies, 130, 131
Mutation, 24, 49
  deleterious, 31
  rate, 30, 35
  recurrent, 28
  –selection balance, 57, 58, 133
Mutual separability, 109

National Forests, 39
Natural selection, *see* Selection, natural
N_e, *see* Population size, effective
Nelder's designs, 115, 116
New Zealand, 58
Noninteractive genotypes, 109–111
Nonrandom mating, 31–36
Nurseries, Wisconsin, 38

Oil content, 47, 48
One-locus model, 12–14, 21, 36
Ortet, 54
Overdominance, 13, 17, 31, 56, 58, 68, 125

Pair lines, 29
Pair mating, 28, 29
Partially Balanced Incomplete Blocks
      (PBIBs), 144
Phenotype, 5, 13, 22, 24, 49, 74
Phenotypic values, 30
Picea abies, 55, 105
Pine
  jack, 38, 41, 71, 101
  loblolly, *see Pinus taeda*
  longleaf, *see Pinus palustris*
  red, *see Pinus resinosa*
  white, 38
Pinus
  caribaea, 110
  palustris, 3
  radiata, 58
  resinosa, 3, 38
  rigida, 69, 70
  taeda, 69, 70, 78, 126, 139, 148
Pleitropy, 84, 87–89
Plus tree, 2, 44

Poisson distribution, 32
Pollen, 42
  contamination, 36, 44
Pollination
  controlled, 34, 43, 59
  open, 59, 78
  supplemental, 34
Polymorphism, 17, 35
Ponderosa pine cover type, 39
Populations
  closed, 63
  copy, 70
  genetics, 158
  introduction of, 133, 134
  multiple, 47, 97, 98, 114, 124, 154
  subdivisions, 51, 97, 125
Population size, 5, 20, 25–31, 46, 47, 49,
    55, 63, 156
  effective (N_e), 26, 33, 34, 42, 49, 60–62,
    64, 77, 78, 92, 102, 115, 131
  finite, 32, 35, 36, 60
  limited, 25, 26, 35
Populus deltoides, 147
Position effects, 20
Principal component analysis, 137
Probability density, 35
  function, 136
Probability of ultimately fixing the favored
    allele (UPF), 26, 27
Progeny population, 23, 32, 59, 61
Progeny tests, 2, 42, 44, 45, 59, 77–79
Provenance, 2
  interprovenance hybrids, 57, 69, 140
  selection, 92, 140
  testing, 50, 134–148

Qualitative traits, 49
Quantitative gene loci, 27
Quantitative trait, 6, 7, 47, 49, 57

Random mating, 16
Rank, 116
  change of, 112, 113
Recessive lethals, 30
Recombinants, 24, 25, 147
Recombination rate, 16, 21, 49
Recurrent Selection (RS), 37–73, 92, 122,
    126, 147, 148

  modified reciprocal (mod. RRS), 52, 67, 68
  reciprocal (RRS), 66–70, 99, 122, 125
  simple (SRS), 43–45, 51, 52, 54, 56, 63,
    64, 74, 99, 121, 125, 131
Regeneration, 42
  artificial, 39
  natural, 8, 39, 41
Regression, 136
  coefficients, 138, 142
  multiple, 135, 141
  multivariate, 136
  nonlinear, 138
Replicate population breeding system, 70,
    112, 127
Republic of Korea, 92
Resistance, 8, 12
  vertical, 12
Response functions, 105–113, 148
  density, 116
  linear, 105–107
  multiparameter, 107
  nonlinear, 107–111
  phenotypic, 108
Response surface, 138–141
Ribonucleic acid, *see* RNA
RNA (ribonucleic acid), 24
Rooting capacity, 55

Salix, 59, 70
Seed
  contamination, 36
  orchard, 2, 8, 20, 34, 41, 43–45, 53, 59,
    61, 72, 78
  production, 34, 42, 44, 72, 154
  zones, 113
Selection, 19, 23, 25–31, 41, 153
  artificial, 3, 5, 8, 9
  cycle of, 42, 56, 59, 61, 74
  density-dependent, 19
  differential, 23, 24, 48, 49, 75, 77, 93,
    146
  direct, 77, 78
  directional, 17, 22, 25, 30, 31
  family, 38, 59, 61, 62, 68, 148
  group, 118
  independent culling level (ICL), 68, 74,
    91–93, 123, 124
  index, 71, 91, 93, 94, 100, 113, 123
  indirect, 79–91

intensity, 28–30, 78
juvenile, 74, 83, 84, 91, 122
limits to, 48–50
mass, 29, 42–44, 47, 56
multiple-index, 124
    selection strategy (MISS), 124, 128
multiple trait-objective, 74, 80, 88, 91–
    94, 97–102, 124, 125
natural, 3, 18, 19
negative, 27
positive, 27, 30
pressure, 24, 25, 27, 30
proportion, 23
recurrent, *see* Recurrent selection
response plateau, 24
response to, 4, 24, 27, 28, 49, 79
tandem, 74, 91–93, 98, 99, 125
within family, 63
Selfing, 31, 58, 59, 122
Separability of environmental effects from
    genotypic, 108
Separability of genotypic effects from envi-
    ronmental, 108
Sexual propagation, 53, 55
Sib mating, 31, 58
Silviculture, 112, 119
Southern pine beetle, 16
Southwide Pine Seed Source Studies, 105
Spacing, 115, 116
Species, 4
    introduction of, 132–134
    testing, 50
Specific combining ability, *see* Combining
    ability, specific
Sublines, 72, 127
Sweden, 59, 70

Tester parent, 60, 64, 66, 102
Tissue culture, 41, 54
Tobacco, 53, 64, 112
    Mosaic Virus, 12
Transcription, 24
Transformations, 109
Translation, 24

Underdominance, 17
Univariate statistics, 135, 143
UPF, *see* Probability of ultimately fixing the
    favored allele

Value function, 140–144
Value surface, 145, 146
Variance: *see also* Variation
    components, 59
    effective number, 33, 60, 61
Variation, *see also* Genetic variation
    heritable, 5
    phenotypic, 74, 77, 78
    response to, 105–111
Vegetative propagation, 3, 41, 53–55

Wahlund effect, 32
White spruce, 71
Wisconsin Department of Natural Resources,
    38
Wood specific gravity, 95

Zones (zonation), 112–115, 125, 131, 148
    environmental, 114, 125

Related Titles Available from Springer-Verlag

Trees
Structure and Function
Editor-in-Chief: H. Ziegler
Journal
ISSN 0931-1890
Trees publishes original articles and reviews on selected topics, treating physiology, biochemistry, functional anatomy, structure and ecology of trees and other woody plants, as well as papers concerned with pathology and technological problems contributing to an understanding of structure and function of trees.

Biotechnology in Agriculture and Forestry
Editor: Yasphal P. Bajaj
With contributions by numerous experts.

Volume 1: **Trees I**
1986. 150 figures. XV, 515 pages. Hard. ISBN 0-387-15581-3.

Volume 2: **Crops I**
1986. 144 figures. XVIII, 608 pages. Hard. ISBN 0-387-15842-1.

Volume 5: **Trees II**
1988. 225 figures. approx. 595 pages. Hard. ISBN 0-387-19158-5.

Volume 6: **Crops II**
1988. 155 figures. approx. 550 pages. Hard. ISBN 0-387-19064-3.

Mechanisms of Woody Plant Defenses Against Insects
Editors: W.J. Mattson, J. Levieux, and C. Bernard-Dagan
With contributions by numerous experts.
1988. 106 figures. XV, 416 pages. Hard. ISBN 0-387-96673-0.

Monographs on Theoretical and Applied Genetics
Coordinating Editor: R. Frankel

Volume 10: **Male Sterility in Higher Plants**
M.L.H. Kaul
1987. 140 figs. 205 tabs. 960 pages. Hard. ISBN 0-387-17952-6.

Plant Gene Research
Managing Editor: Th. Hohn

Volume 3: **A Genetic Approach to Plant Biochemistry**
Editors: A.D. Blonstein and P.J. King
With contributions by numerous experts.
1986. 30 figures. XI, 291 pages. Hard. ISBN 0-387-81912-6.

Volume 5: **Temporal and Spatial Regulation of Plant Genes**
Editors: D.P.S. Verma and R.B. Goldberg
With contributions by numerous experts.
1988. approx. 340 pages. Hard. ISBN 0-387-82046-9.

Springer Series in Wood Science
Editor: T.E. Timell

Vascular Differentiation and Plant Growth Regulators
L.W. Roberts, P.B. Gahan, and R. Aloni
1988. 28 figures. X, 154 pages. Hard. ISBN 0-387-18989-0.

Sampling Theory for Forest Inventory
Pieter G. de Vries
1986. 20 figures. X, 399 pages. Soft. ISBN 0-387-17066-9.